Pierre-Yves Mantel

Molecular mechanisms of T cell tolerance

Pierre-Yves Mantel

Molecular mechanisms of T cell tolerance

Regualtion of FOXP3 expression: a key transcription factor for regulatory T cells

Südwestdeutscher Verlag für Hochschulschriften

Impressum/Imprint (nur für Deutschland/ only for Germany)
Bibliografische Information der Deutschen Nationalbibliothek: Die Deutsche Nationalbibliothek verzeichnet diese Publikation in der Deutschen Nationalbibliografie; detaillierte bibliografische Daten sind im Internet über http://dnb.d-nb.de abrufbar.
Alle in diesem Buch genannten Marken und Produktnamen unterliegen warenzeichen-, marken- oder patentrechtlichem Schutz bzw. sind Warenzeichen oder eingetragene Warenzeichen der jeweiligen Inhaber. Die Wiedergabe von Marken, Produktnamen, Gebrauchsnamen, Handelsnamen, Warenbezeichnungen u.s.w. in diesem Werk berechtigt auch ohne besondere Kennzeichnung nicht zu der Annahme, dass solche Namen im Sinne der Warenzeichen- und Markenschutzgesetzgebung als frei zu betrachten wären und daher von jedermann benutzt werden dürften.

Verlag: Südwestdeutscher Verlag für Hochschulschriften Aktiengesellschaft & Co. KG
Dudweiler Landstr. 99, 66123 Saarbrücken, Deutschland
Telefon +49 681 37 20 271-1, Telefax +49 681 37 20 271-0, Email: info@svh-verlag.de
Zugl.: Universitaet Zuerich. Life Sciences, Diss., 2007

Herstellung in Deutschland:
Schaltungsdienst Lange o.H.G., Berlin
Books on Demand GmbH, Norderstedt
Reha GmbH, Saarbrücken
Amazon Distribution GmbH, Leipzig
ISBN: 978-3-8381-0536-9

Imprint (only for USA, GB)
Bibliographic information published by the Deutsche Nationalbibliothek: The Deutsche Nationalbibliothek lists this publication in the Deutsche Nationalbibliografie; detailed bibliographic data are available in the Internet at http://dnb.d-nb.de.
Any brand names and product names mentioned in this book are subject to trademark, brand or patent protection and are trademarks or registered trademarks of their respective holders. The use of brand names, product names, common names, trade names, product descriptions etc. even without a particular marking in this works is in no way to be construed to mean that such names may be regarded as unrestricted in respect of trademark and brand protection legislation and could thus be used by anyone.

Publisher:
Südwestdeutscher Verlag für Hochschulschriften Aktiengesellschaft & Co. KG
Dudweiler Landstr. 99, 66123 Saarbrücken, Germany
Phone +49 681 37 20 271-1, Fax +49 681 37 20 271-0, Email: info@svh-verlag.de

Copyright © 2009 by the author and Südwestdeutscher Verlag für Hochschulschriften Aktiengesellschaft & Co. KG and licensors
All rights reserved. Saarbrücken 2009

Printed in the U.S.A.
Printed in the U.K. by (see last page)
ISBN: 978-3-8381-0536-9

Table of Contents

Summary 3

Zusammenfassung 5

1. Introduction 9

    1.1. Overview 9

    1.2. The innate immune system tailors the adaptive immune response 10

    1.2. Mechanisms of Th2 cells differentiation 14

        1.2.1. GATA3 is a master regulator of Th2 cells differentiation 16

        1.2.2. Flexibility of commitment 18

    1.3. Immune tolerance 19

        1.3.1. Central tolerance 19

        1.3.2. Peripheral tolerance 20

    1.4. $CD4^+CD25^+FOXP3^+$ T regulatory cells 21

        1.4.1. Tregs of thymic origin 23

        1.4.2. Peripheral generation of $CD4^+CD25^+FOXP3^+$ Tregs 26

        1.4.3. Cytokines involved in the generation and maintenance of Tregs 27

        1.4.4. FOXP3 29

    1.5. Concluding remarks and aim of the study 31

2. Results 32

    2.1. Molecular mechanisms underlying FOXP3 induction in human T cells 32

    2.2. GATA3 driven Th2 responses inhibit FOXP3 expression and the formation of regulatory T cells 59

    2.3 Statement of contribution to publications 91

3. Discussion 92

    3.1. Identification and characterization of the human basal FOXP3 promoter 92

    3.2. Cell-specific activity of the FOXP3 promoter 93

    3.3. Effects of immunosuppressive drugs on FOXP3 expression 94

    3.4. Models of Treg response to antigen 95

3.5. Differentiating Th2 cells lack FOXP3 96

3.6. GATA3 inhibits FOXP3 expression 97

3.7. GATA3 represses the FOXP3 promoter activity 99

3.8. Effect of IL-4 on committed Tregs 99

3.9. Effect of IL-4 on Tregs in vivo 100

3.10. Conclusion and outlook 101

4. References *103*

ACKNOWLEDGEMENTS *136*

Summary

The T regulatory cells (Tregs) play an important role in immune homeostasis, by maintaining tolerance to self-antigens and allergens as well as by limiting inflammatory tissue damage during chronic infections. Humans or mice lacking the Treg-associated transcription factor FOXP3 develop severe features of autoimmunity and allergy. Ectopic FOXP3 expression endows non-regulatory T cells with many of the hallmarks of regulatory T cells and transgenic reporter mice carrying a FOXP3-GFP indicated that FOXP3 expression correlates with suppressive T cells. Therefore FOXP3 is a faithful marker for Tregs.

Although Tregs can be generated in the thymus, their induction out of non-Tregs was also revealed in the periphery. The molecular mechanisms leading to Tregs generation in the periphery are still not identified. To gain insight into this key process we analyzed peripheral FOXP3 regulation. For this purpose we localized and cloned the human FOXP3 promoter into a reporter plasmid and characterized its activity in primary $CD4^+$ T cells. In a second step, we looked at factors, which regulate FOXP3 expression during the differentiation process from naïve T cells into the effectors cells.

This thesis reveals that the FOXP3 promoter is located -6221 bp upstream of the translation start site and the 5' UTR is interrupted by a 6000 bp intron. Our results indicate that the FOXP3 promoter is cell-specific and is active only in primary CD4+ T cells.

We demonstrated that FOXP3 mRNA and protein expression is induced following TCR engagement in $CD4^+CD25^-$ T cells or artificially by phorbol 12-myristate 13-acetate (PMA) and ionomycin induced FOXP3 promoter activity. The activation-responsive element of the FOXP3 promoter is composed of at least three NFAT and AP-1 sites.

Cyclosporin A (CsA) completely inhibited the mRNA induction of FOXP3 as well as the promoter activity. CsA is a well-known immunosuppressive drug, which blocks NFAT translocation into the nucleus by inhibition of the calcineurin phosphatase activity. We have shown that the immunosuppressants glucocorticoids and rapamycin promote FOXP3 expression. Therefore immunosuppressive drugs may have different mechanisms to promote immune tolerance and a more precise knowledge of the immunosuppressive drugs targets will improve their therapeutical usage.

Although induction of FOXP3 upon TCR triggering is probably an important step, it remains unclear, which signals are specifying Tregs induction over the induction of effector Th1 or Th2 cells.

We demonstrated that Th2 commitment prevents the induction of $FOXP3^+$ Treg cells by a GATA3-dependent mechanism., in contrast to naïve T cells cannot *in vitro* differentiated Th2 cells express FOXP3 upon stimulation with TGF-β. The Th2 cytokine IL-4 efficiently inhibited FOXP3 mRNA and protein expression in differentiating naïve human T cells.

We investigated the role of GATA3 in this process, since it is known to be essential for Th2 commitment and is induced by IL-4. Transient overexpression of GATA3 blocked the induction of the FOXP3 promoter activity in human T cells and strikingly, mice engineered to overexpress GATA3 in T cells (CD2-GATA3 x DO11.10 Tg mice) do not express FOXP3 following TGF-β exposure along with the specific antigen. We discovered that these GATA3 effects are mediated by a direct action on the FOXP3 promoter. We described a GATA3 binding site in the FOXP3 promoter, between the NFAT-inducible region and the transcription start site. Site-specific mutation of this GATA3 binding region, which is accessible on the chromatin level, reveals that this element negatively regulates the FOXP3 promoter activity.

Taken together this thesis revealed that antigen-experience is an important step in Treg generation. We identified GATA3 as a negative regulator of FOXP3 expression, which suggests that Treg induction relative to Th1 or Th2 differentiation is a matter of negative cross-regulation of competing lineage-specific factors. This mechanism is likely to improve our understanding of Treg induction and thus the induction of immune tolerance in disease such as allergy or autoimmunity.

Zusammenfassung

Regulatorische T-Zellen (Tregs) halten die Immuntoleranz gegenüber Selbstantigen oder Allergen aufrecht, in dem sie die entzündliche Gewebezerstörung begrenzen. Die Tregs sind durch die Expression von FOXP3 gekennzeichnet, welches im Falle von genetischen Defekten zu schweren Autoimmunerkrankungen und allergischer Entzündungen führt. Artifizielle FOXP3-Überexprimierung verleiht normalen T Zellen das phänotypische Erscheinungsbild von Tregs. Studien über transgene Reporter Mäuse (FOXP3-GFP) zeigen, dass die FOXP3 Expression mit aktiven suppressor-T-Zellen korreliert. Dem zufolge ist FOXP3 ein wichtiger Indikator für Tregs, welche in der Regel im Thymus generiert werden. Interessanterweise konnte kürzlich gezeigt werden, dass Tregs auch in der Peripherie induziert werden können und somit auch bei der Toleranz von Allergenen wichtig sein könnten.

Die molekularen Mechanismen der Treg Induktion in der Peripherie sind noch nicht erforscht worden. Gegenstand dieser Studie ist daher der Prozess der peripheren Treg Induktion. Das FOXP3-Gen wurde als wichtige molekulare Leitstruktur verfolgt. Dementsprechend wurde der humane FOXP3 Promoter lokalisiert und in einen Luciferase-Vektor kloniert, um dessen Aktivität in primären humanen T-Zellen zu bestimmen. Ausserdem wurden Faktoren lokalisiert, die die FOXP3-Expression während der T-Zell-Differenzierung regulieren.

Die vorliegenden Ergebnisse zeigen, dass der FOXP3-Promoter -6221 bp aufwärts von der Translation Start-Stelle liegt und zu der 5'-UTR durch ein 6000 bp Intron unterbrochen ist. Die Studie beschreibt ausserdem die Zell-spezifische Aktivität des FOXP3-Promoters, die sich nur in primären CD4+ T-Zellen darstellen lässt.

Die FOXP3 mRNA- und Proteinexpression wird durch die T-Zell-Rezeptor (TCR) Aktivierung oder artifiziell durch phorbol 12-myristate 13-acetate (PMA) und Ionomycin induziert. Die aktivierungsabhängigen Elemente des FOXP3-Promoters sind aus mindestens drei NFAT und mehrere AP-1 Bindungsstellen aufgebaut.

Cyclosporin A (CsA) inhibiert sowohl die FOXP3 mRNA- und Proteinexpression, als auch die FOXP3-Promotoraktivität. CsA ist ein weitverbreitetes, immunsupressives Medikament, dessen Wirkung auf der Blockierung des NFAT beruht. Es §wird deutlich, das immunsupressive Medikamente in unterschiedlicher Art und Weise die Immuntoleranz über Tregs beeinflussen kann. Das Verständnis des Wirkmechanismus bezüglich des FOXP3 und der Treg-Induktion wird den Einsatz dieser Medikamente hinsichtlich der Immuntoleranz verändern und die therapeutischen Anwendungen erweitern.

Obwohl die Induktion des FOXP3 bzw. der Tregs von dem TCR abhängt, bleibt die Spezifität der Treg-Induktion gegenüber den Effektor Th1 oder Th2 Zellen unklar. In der vorliegenden Arbeit zeigen wir erstmals, dass die Th2-Differenzierung die Induktion von FOXP3+ Tregs über einen GATA-3 abhängigen Mechanismus verhindert. In Gegensatz zu naïven T-Zellen können *in vitro* differenzierte Th2-Zellen kein FOXP3 exprimieren.

Bei unserer Analyse des zugrundeliegenden Mechanismus konnten wir die wichtige Rolle des GATA-3 beweisen, welches für die Th2-Differenzierung unersetzlich ist und durch IL-4 induziert wird. Die transiente Überexpression von GATA-3 blockiert die Induktion der FOXP3-Promoter Aktivität in humanen T-Zellen. In Mäusen, die genetisch so verändert wurden, dass sie GATA-3 nur in T-Zellen exprimieren, konnte im Gegensatz zu den Wild-Typ Mäusen keine FOXP3-Induktion nach TGF-β- und Antigen-Stimulation beobachten werden. Es konnte bewiesen werden, dass GATA-3 an eine Stelle im FOXP3-Promoter bindet, die zwischen der NFAT-induzierbaren Region und der Transkriptions-Start-Stelle liegt. Die im Chromatin zugängliche GATA-3 Bindestelle inhibiert die FOXP3-Expression, wie es durch die Sequenz-spezifische Mutation bewiesen werden konnte.

Zusammenfassend lässt sich auf Grund der vorliegend Doktorarbeit feststellen, dass Antigen-Kontakt eine Schlüsselrolle in der Treg-Induktion spielt. Die GATA-3 vermittelte negative Regulation des FOXP3-Gens lässt vermuten, dass die Spezifität der Treg-Induktion durch negative Regulations-Mechanismen zustande kommt. Dieser Mechanismus ist wichtig für die immunologischen Konzepte der Treg- und Toleranzinduktion bei Allergien und Autoimmunerkankungen.

Abbreviations

APC	Antigen-Presenting cells
AIDS	Aquired immunodeficiency syndrome
BTEB1	Basic transcription element binding protein 1
CD	cluster of differentiation
CsA	Cyclosporin A
CTL	Cytotoxic T cells
CTLA-4	Cytotoxic T lymphocyte antigen 4
DC	Dendritic cell
DNMT	DNA methyltransferase
EAE	Experimental autoimmune encephalomyelitis
FOG	Friend of GATA
GITR	Glucocorticoid-induced tumor necrosis factor receptor family-related receptor
GVDH	Graft-versus host disease
IFN	Interferon
IL	Interleukin
GPI	Glycosylphosphatidylinositol
LPS	Lipopolysaccharide
MAPK	Mitogen-activated protein kinase
MHC	Major histocompatibility complex
MBD2	methyl-CpG binding domain protein-2
MBP	Myelin basic protein
MS	Multiple sclerosis
NFAT	Nuclear factor of activated T cells
NuRD	nucleosome remodeling and deacetylase
OVA	Ovalbumin
PAMP	Pathogen-associated molecular patterns
PD-1	Programmed death-1
PMA	Phorbol 12-myristate 13-acetate
ROG	Repressor of GATA
STAT6	signal transducer and activator of transcription
TCR	T cell receptor

TGF-β	Transforming growth factor-β
Th	T helper cell
TLRs	Toll-like receptors
Tr1	T regulatory 1
TSLP	Thymic stromal lymphopoetin
TSS	Translation start site
TIEG2	TGF-β-inducible early protein gene 2

1. Introduction

1.1. Overview

The human organism is continuously exposed to microorganisms, which constitute a potential danger to the host. The immune system, which fights against pathogens, has the difficult task to distinguish between harmless and harmful antigens. Although the protection against infections is fundamental for the survival of all animals, an immune reaction to self or innocuous antigens may cause severe disease leading to autoimmunity or allergies with in some cases fatal consequences. Therefore the immune system has developed an efficient barrier to identify pathological microorganisms and to actively maintain tolerance to innocuous antigens. The immune system is divided into two parts: innate and adaptive.

The innate immune system is the first line of defense against pathogenic microorganisms (bacteria, viruses, fungi, and parasites). Once activated, the cells produce cytokines that regulate and coordinate many activities of the cells from the innate and adaptive immune system. Innate immunity is activated by specific structures, which are common to different microbes but does not allow an efficient defense against the invading pathogens, which are structurally variable and is therefore complemented by the adaptive immune system. The adaptive immunity may take days or weeks, after an initial infection, to have an effect. It is composed of humoral and cell-mediated immunity including B cell, CD8 cytotoxic T cells (CTL) and the effector or T helper cells (Th). The T helper cells produce cytokines that activate macrophages and induce proliferation of B and T cells. Innate and adaptive systems communicate to build an efficient way of fighting pathogens.

1.2. The innate immune system tailors the adaptive immune response

The specific immune response is driven by effector cells composed of the Th1, Th2 and Th17 cells. The Th1/Th2/Th17 response is further controlled by T regulatory cells (Tregs), which are potent suppressors of the immune system. The immune response is a dynamic process, which has to be adapted for the specific invading pathogens, in order to efficiently clear the infection. In addition the immune response may tend toward tolerance, instead of activation and elimination of the microbes. Therefore, an accurate communication has developed between the innate immune system, which encounters first the pathogen and the adaptive immune system, which fights specifically and efficiently against antigens.

The "innate" receptors, Toll-like receptors (TLRs), mannose receptor recognize conserved pathogenic particles called pathogen-associated molecular patterns (PAMPs) expressed by microbes such as lipopolysaccharide (LPS) and triggers a type 1 response (Th1 cells) characterized by a high IFN-γ secretion. For example, viruses, which infect the cells, will be recognized by the intracellular TLRs, which bind to the ssDNA or dsRNA typical from viruses. HSV-2 triggers a potent type I interferon response by activating the TLR9 from the plasmacytoid dendritic cells (DCs). This interferon secretion leads to the anti-viral state, characterized by inhibition of viral replication, increase effectiveness of CTL-mediated killing of infected cells and stimulation of Th1 cell development [1-4]. Intracellular bacteria are engulfed by phagocytes, in which they can survive, but during this intracellular persistence, the pathogens are degraded and presented to the $CD4^+$ T cells, which in turn secrete IFN-γ increasing the phagocytic activity of the phagocytes and lead to the elimination of the bacteria. Furthermore macrophages secrete IL-12, which induces differentiation of naïve T cell into the IFN-γ-secreting Th1 cells [5]. Therefore a Th1 response is generated in response to infections by viruses or intracellular bacteria. The critical role of T cells is demonstrated by patients suffering from acquired immunodeficiency syndrome (AIDS), who are extremely susceptible to infections by intracellular bacteria [6,7] and further illustrates that innate and adaptive immune systems collaborate in order to fight efficiently against pathogens [8,9]. Although this interconnection between innate immunity is better understood in the generation of a type 1 immune response, innate receptors have also been reported to efficiently generate a type 2 immune response, characterized by Th2 cells, which secrete the cytokine IL-4. Th2 cells are fighting helminth infections by modulating the antibody response. Allergy is an immune disorder characterized by an exuberant Th2 cell activity, resulting in a switch from IgM to IgE production in B cells [10]. Allergens are proteins, often enzymes, and are normally harmless, but provoke a reaction in allergic patients already at low concentrations (ng - µg range) [10]. IL-4 is the most potent factor inducing differentiation into Th2 cells *in vitro*. The origin of the Th2-driving forces *in vivo* are not clear, since several cells secrete IL-4 in basal conditions: Th2, basophiles, mast cells and eosinophils. After differentiation, Th2 cells secrete IL-4. This effect would propagate as CD4+ effector cells are differentiating at the antigen presentation site, leading to a threshold in IL-4, which drives the expression of the Th2 profile. The pathogens, which are first recognized by the cells of the innate immune system, influence the maturation stage of APCs and thus the strength of the TCR signal and costimulation that receive the T cells. In addition pathogens affect the cytokines produced by the innate system and therefore the immune polarization and differentiation into the different effector lineages.

The naïve T cells require 3 signals in order to differentiate into effector cells, TCR stimulation, costimulation and cytokines. The first signal is the TCR stimulation by the peptide presented on the major histocompatibility complex class II (MHCII) of an antigen-presenting cell (APC), only T cells with a TCR specific for the antigen-MHC II complex will be activated, providing the antigen-specificity of the immune response. Depending on its strength, TCR stimulation influences the differentiation of cells. Strong stimulation leads to Th1 commitment, while weak antigen stimulation favors differentiation into Th2 [11,12]. The second signal is given by costimulation, which is an important stimulus, since a lack of this signal leads to anergy (unresponsive state). The costimulatory molecules can be divided into activators or inhibitors of the immune response: CD28, ICOS stimulation potently induces the T cell activation and cytokine secretion whereas cytotoxic T lymphocyte antigen 4 (CTLA-4), PD1 act as inhibitory molecules [13-20]. The third signal is given by the cytokines secreted by cells of the innate immune system or in an autocrine fashion.

The type I polarizing factors are IL-12 [5,21], IL-23 [22] and IL-27 [23,24]. IL-12, IL-23 and Il-27 all belong to the same cytokine family [25] and are mostly produced by activated monocytes, macrophages, neutrophils and dendritic cells. These cytokines induce STAT1 [26,27], STAT4 [28] and T-bet [29], which are triggering the production of IFN-γ [30], the typical Th1 cytokine.

The type 2 polarizing factors are mainly IL-4 and notch ligands [31-36] that lead to the induction of the IL-4-secreting Th2 cells. In addition, tolerance might be induced by modulation of the DCs function. Induction of dendritic cells maturation by pathogens is a key process in naïve T cell differentiation into Th1 and Th2 cells. In the steady state, dendritic cells are immature and provide a tolerogenic environment. Key cytokines for converting dendritic cell into more professional tolerogenic cells are IL-10 and TGF-β [37].

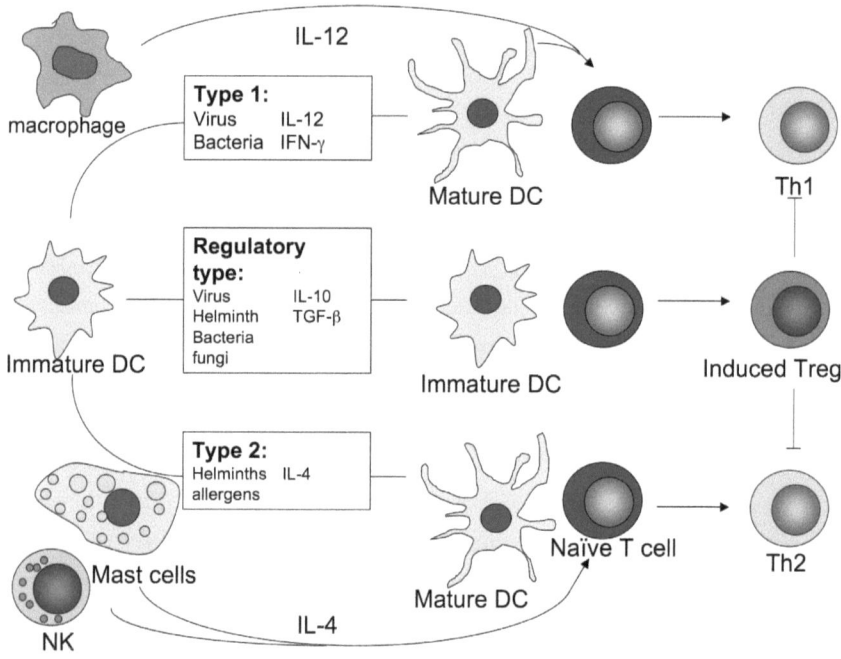

Figure 1: Polarization of the adaptive response by the innate immune system. The pathogens can be categorized into type 1, type 2 or tolerizing pathogens depending on the response induced. The pathogens affect maturation of the DCs or other APCs and cytokine secretion by the cells of the innate immune system, inducing the differentiation of antigen-specific naïve T cells into different subsets, adapted from [38].

Several pathogens take advantage of this fundamental strategy of the immune system and have developed strategies to escape the immune system by inducing regulatory DCs. Regulatory or tolerogenic DCs are characterized by reduced expression of costimulatory molecules in particular CD40, CD80, CD86, reduced IL-12 and increased IL-10 production [39].

IL-10 has been shown *in vitro* to induce differentiation of immature DC into tolerogenic DC, which promote the generation of a subset of IL-10-secreting regulatory T cells (Tr1) [40]. Plasmodium falciparum [41,42], Mycobacteria [43,44], hepatitis C [45], herpes simplex virus [46], cytomegalovirus [47], S. mansoni [48], bordetella pertussis [49] induce tolerizing dendritic cells, giving rise to tolerance and chronic infection. Therefore differentiation of effector cells from naïve cells is a complex process occurring in a very dynamic environment, under the scrutiny of the innate immune system, which tailors the adaptive response to generate a strongly adapted and efficient response to the invading pathogens (Figure 1).

The Th1 and Th2 cells were discovered first by Mosmann. He suggested that the immune system response to pathogens involves the production of two clusters of cytokines with antagonistic effect for the other subset [50-52]. Each cell subset secretes a pattern of cytokines determining the functional diversity of $CD4^+$ memory T cells and type of induced immune response.

The ability of mounting a Th1 or Th2 response is strongly under control of genetic factors as illustrated by the Th1-prone C57BL/6 and the Th2-prone Balb/C mice strains. The Balb/C mice are highly susceptible to induced asthma and inefficient in mounting a Th1 immune response to the intracellular parasite Leishmania major [53]. On the other hand, C57BL/6 are not susceptible to asthma induction and clear efficiently infections by L. major [54,55].

In addition, environmental factors play an important role. In the last twenty years, the incidence of atopy in developed countries increased dramatically [56], probably as a consequence of the decrease of infections with type I polarizing pathogens during childhood due to improved hygiene in the industrialized world [56]. However, the role of Th1 responses downregulating Th2 responses is still controversial since atopic children infected with influenza virus exacerbate the symptoms of asthma. Thus, respiratory viral infection and the acute Th1 response can positively regulate Th2-dependent allergic pulmonary disease *in vivo*, at least in part [57], indicating that other cells might be involved in the regulation of the Th1/Th2 balance. Indeed T regulatory cells with potent *in vitro* and *in vivo* suppressive capacity regulate the Th1 and Th2 cells [58-61] and control the T cell response. The Th17 cells belong to another cell lineage, which produces IL-17. The discovery of Th17 and inducible Treg, which can differentiate from the same precursor cell, the naïve T cell, further completes the Th1/Th2 paradigm. At least four different T cell lineages, characterized by their cytokine profile and functions, can differentiate from the naïve T cells. Treatment of naïve T cells with TGF-β induces cells with regulatory properties (Tregs), which are able to suppress activation, proliferation and cytokine production of CD4 cells. Tregs are characterized by their suppressive function and constitutive expression of CD25 and FOXP3 [62-64]. Interestingly when IL-6 is added to TGF-β at the time of priming, the induction of Treg is inhibited leading to the generation of Th17 cells [65-67], which are characterized by a potent pro-inflammatory capacity. These cells have been described in several autoimmune conditions

[65,68,69]. However, the molecular mechanisms involved in Treg and Th17 cells generation are still not known (Figure 2).

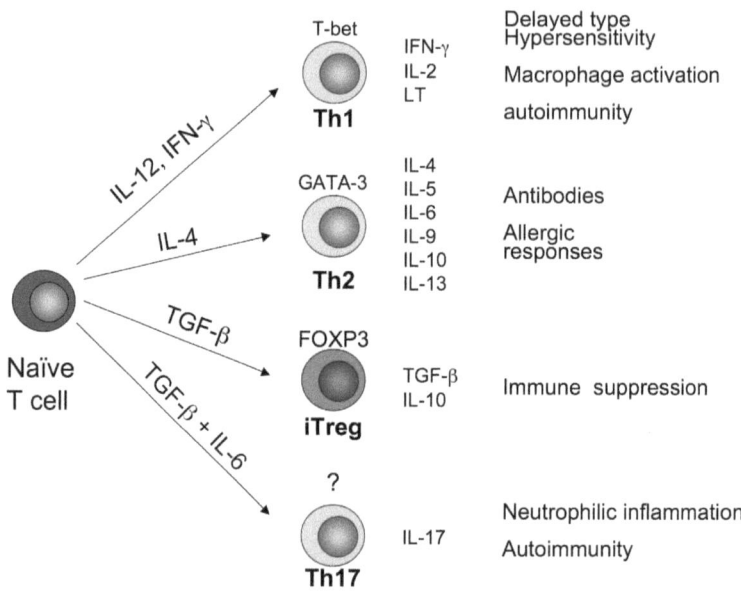

Figure 2: Different subsets of T effector cells can be generated out of naïve T cells. The cytokines present at the time of priming of the naïve cells lead to Th1, Th2, iTregs and Th17. Although they originate from the same precursors the cells have very different functions.

1.2. Mechanisms of Th2 cells differentiation

Type 2 responses are characteristic of the beneficial immune response to helminth parasitic infection, but also of the inappropriate immune response leading to allergy and asthma [70], graft-versus host disease (GVDH), progressive systemic sclerosis, systemic lupus erythematosus [71]. Il-4 is a potent Th2-driving cytokine *in vitro*, but the mechanisms of Th2 cells differentiation *in vivo* are in contrast to Th1 cells far from being fully understood. The cellular sources of IL-4 during an allergic reaction or helminth infections are not well-defined and may depend on the localization and the antigen type. Mast cells, basophils, NKT cells and previously differentiated Th2 cells display high basal levels of IL-4 mRNA and can rapidly release IL-4 upon stimulation. Indeed, it has been shown that IL-4 production during *Nippostrongylus Brasiliensis* infection develops independently of the adaptive immune system, but comes rather from cells of the innate immune system. IL-4 production by Th2 cells is then more important for the effector phase to maintain a type 2 polarized immune response [72-74].

Th2 cells are characterized by the production of the closely related IL-4, IL-5 and IL-13, whereas these genes are silenced in Th1 cells [75]. They are located in the same cluster on chromosome 5 in humans [76] and 11 in mice [76]. The initiation phase of the differentiation process occurs as the TCR is triggered by antigen and consists of a complex array of epigenetic changes and transcription factor expression and induction, which leads to the characteristic cytokine profile expression.

Naïve T cells express the IL-4 receptor, a heterodimer composed of the specific IL-4Rα subunit and the common γ-subunit [77]. Upon binding to its receptor, IL-4 can initiate the phosphorylation and activation of Stat6. Activated Stat6 will form a dimer, which enables it to enter the nucleus where it will, together with NFAT, AP-1 NF-κB and other TCR-induced signals activate the transcription of GATA3 and IL-4 [78,79]. Autocrine IL-4 production per se reinforces Th2 differentiation.

Thus, Stat6 is a central mediator of the IL-4 signal involved in Th2 development [80,81]. The importance of IL-4 and Stat6 in Th2 differentiation has been demonstrated by the generation of IL-4 or Stat6 target-specific deficient mice. CD4 T cells from IL-4-deficient mice have a defect to mount a Th2 response or produce IL-5 and IL-13 after *in vivo* challenge [82,83]. Similarly, in Stat6-deficient mice, Th2 differentiation was blocked, and IgE titers in response to *N. brasiliensis* are dramatically decreased [81,84].

Stat6 is involved in direct regulation of GATA3 expression, a transcription factor [85], which also increases Th2 cell differentiation.

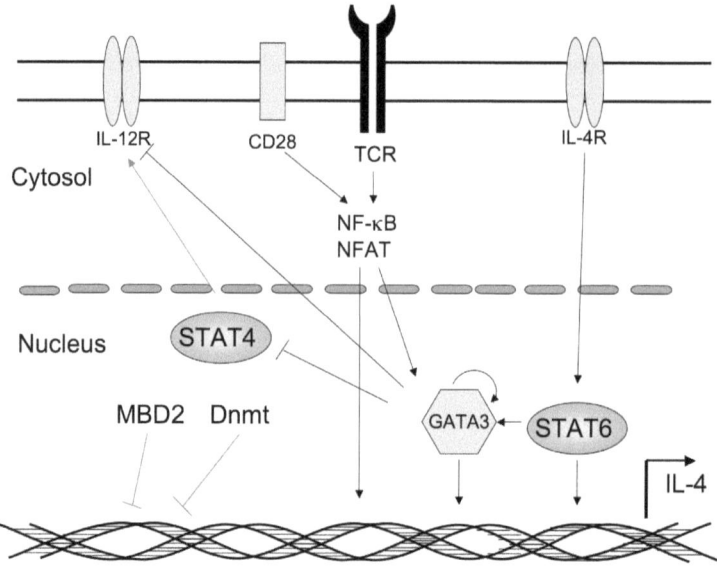

Figure 3: Induction of Th2 cell differentiation. IL-4, TCR signaling and costimulation trigger Th2 cell differentation by inducing STAT-6, NFAT, NFκB and GATA3. These transcription factors collaborate to mediate chromatin remodeling of the IL-4, IL-5, IL-13 locus and activate gene transcription.

1.2.1. GATA3 is a master regulator of Th2 cells differentiation

Naïve T cells express a basal level of GATA3, which is upregulated in the course of Th2 differentiation and extinguished during Th1 differentiation [86]. GATA3 expression in primary T cell is dependent on TCR activation and is blocked by the NF-κB inhibitor SN50. In addition, mice that lack the p50 subunit of NF-κB are unable to mount airway eosinophilic inflammation due to the inability of the p50-/- mice to produce IL-4, IL-5 and IL-13: cytokines that play a key role in asthma pathogenesis. CD4$^+$ T cells from p50-/- mice failed to induce GATA3 expression under Th2-differentiating conditions, but showed unimpaired T-bet expression and IFN-γ production under Th1-differentiating conditions. Inhibition of NF-κB activity prevents GATA3 expression and Th2 cytokine production in developing, but not in committed Th2 cells [87,88].

Regulation of cytokine gene expression is strongly mediated by chromatin remodeling characterized by histone acetylation and DNA methylation. GATA3 has been shown to induce chromatin structure remodeling of the Th2 locus IL-4, IL-5, IL-13 and IL-10 allowing access to the transcriptional machinery [75,89,90]. Overexpression of GATA3 in Th cells induces the appearance of the Th2-specific DNase I-hypersensitive sites II, III, and V of the *il4* gene as well as the hyperacetylation and demethylation of the *il4* locus [75,91,92]. In addition, conditional inactivation of the *gata3* gene leads to decreased histone acetylation and increased DNA-methylation of the IL-4 locus [91]. It has been shown that GATA3 inhibits and competes with the binding of the methyl-CpG binding domain protein-2 (MBD2) to the second intron of the *il4* gene and to CNS-1 and thus the recruitment of a silencing complex [93]. MBD2 is capable of recruiting the multiprotein nucleosome remodeling and deacetylase (NuRD) repressive complex [94]. In the MBD2 -/- mice, GATA3 is not necessary for heritable induction of IL-4 and the progeny express substantial levels of IL-4.

$CD4^+$ single-positive thymocytes express IL-4, but attenuate GATA3 expression, and recruit DNA methyltransferases (Dnmts) to the Il4-Il13 locus and downregulate IL-4 expression as they mature into T cells. Type 2 polarization blocks Dnmt1 recruitment, enhances histone H3 Lys4 methylation (indicative of accessible chromatin) and initiates DNA demethylation of the locus. Dnmt1-/- CD4 and CD8 T cells derepress IL-4 expression considerably, demethylate DNA and increase H3 Lys4 methylation without affecting GATA3 expression, demonstrating that Dnmt1 and DNA methylation are essential for proper Il4 regulation. These data indicate that Dnmts, DNA as well as histone methylation, and transcription factors work together in determining appropriate Il4 expression patterns [95].

In addition to its role in remodeling chromatin, GATA3 acts by directly transactivating the promoter of IL-5, IL-13 [96,97] and the enhancer of IL-4 [79,98]. Investigations using transgenic mice containing the murine Th2 cytokine cluster carrying an IL-4 promoter-luciferase reporter showed that IL-4 promoter activity in effector CD4 T cells from these transgenic mice was strong, and importantly Th2 specific. Expression of the IL4 promoter reporter was transactivated *in vivo* by GATA3 [99].

In fact reduction of GATA3 expression in cloned Th2 cells by antisense RNA led to a reduction of IL-4, IL-5, IL-6, IL-10 and IL-13 Th2 cytokine mRNA and protein secretion [100]. Interestingly, GATA3 does not only promote Th2 cells commitment by increasing transcription of Th2 cytokines, but also by repressing Th1 commitment. GATA3 interferes with IL-12 signaling by downregulating STAT4, which is necessary for IFN-γ production [101,102].

The different studies demonstrate that GATA3 is the master transcription factor in Th2 cell differentiation. In addition to GATA3, cMaf and JunB, two bZIP transcription factors are expressed specifically in Th2 cells and bind to the IL-4 proximal promoter [103-106].

Overexpression of c-Maf in non-T cells induced IL-4 mRNA expression. Furthermore, c-Maf-/- mice secrete less IL-4 [107]. JunB, a member of the AP-1 transcription factor family, acts synergically with c-Maf to induce IL-4 expression.

Importantly GATA3 autoactivates its own transcription in a stat-6-independent mechanism, giving rise to a positive feedback that stabilizes and reinforces Th2 commitment [108].

GATA3 expression is not only essential for Th2 cell differentiation but also for maintenance of established chromatin remodeling at the Th2 cytokine gene loci, including Th2-specific long range histone hyperacetylation of the IL-13/IL-4 gene loci. By using a Cre/LoxP-based site-specific recombination system in cultured CD4 T cells, Yamashita et al. investigated the effect of loss of GATA3 expression by *in vitro* differentiated Th2 cells. After ablation of GATA3, the production of all Th2 cytokines was reduced, the DNA methylation at the IL-4 gene locus was increased, and histone hyperacetylation at the IL-5 gene was decreased. Thus, GATA3 plays important roles in the maintenance of the Th2 phenotype and continuous chromatin remodeling of the specific Th2 cytokine gene locus through cell division [90].

1.2.2. Flexibility of commitment

Once a cell has been committed into a certain T cell subset, the phenotype is imprinted and will be inherited by the sister cells. Therefore, $CD4^+$ T cell priming under Th1 or Th2 polarization conditions gives rise to polarized cytokine gene expression. In these conditions, human naive T cells acquired stable histone hyperacetylation at either the *Ifnγ* or *Il4* promoter. But some flexibility is possible, hypoacetylation of the nonexpressed cytokine gene does not lead to irreversible silencing, restimulation of Th1 or Th2 cell clones into Th2 or Th1 conditions resulted in cells producing both IL-4 and IFN-γ. Thus, the chromatin acquires also an open conformation in the previously closed and repressed chromatin. But some cells, from central memory expressing CRTh2, a prostaglandin D2 receptor expressed by some Th2 cells, failed to upregulate T-bet and to express IFN-γ when stimulated under Th1 conditions [109,110]. Thus, most human $CD4^+$ T cells retain both memory and flexibility of cytokine gene expression.

1.3. Immune tolerance

Immunologic tolerance is defined as unresponsiveness to an antigen that is induced by previous exposure to that antigen. When specific lymphocytes encounter antigens, the lymphocytes may be activated, leading to immune responses, or the cells may be inactivated or eliminated, leading to tolerance. In order to avoid an immune response against self-antigen and harmless antigen, several mechanisms have evolved, including deletion of self-reactive cells in the thymus in a process called central tolerance, as well as deletion, anergy and active suppression by Tregs in the periphery.

1.3.1. Central tolerance

The first mechanism to prevent autoimmune reactions is deletion of self-reactive lymphocytes and is called central tolerance, which occurs in the thymus during the maturation of the lymphocytes [111]. The precursors of the T lymphocytes originate from the bone marrow. Then, they migrate to the thymus, where they will go through a complex process of maturation and differentiation to finally become functional T lymphocytes. This maturation is characterized by TCR formation and by exclusive expression of CD4 or CD8 [112]. During the maturation process, the T cell rearrange randomly their TCR [113], creating many variants, which are useless since they are unable to recognize antigen-self-MHC with a high enough affinity. TCRs with too low affinity for self- MHC will die by neglect, and commit apoptosis. This process is called positive selection and it allows that only cells expressing a TCR that can interact with self-peptide–MHC complexes to differentiate further [114].

The hallmark of central tolerance is clonal deletion characterized by suicide of T-cell progenitors that have high affinity for self-antigens. Strongly self-reactive progenitors are under strict control and it is the moderately reactive progenitors that mature, populate the lymphoid organs and participate in immune responses to foreign antigens [111,115,116].

Central tolerance is an efficient process, but some self-reactive cells may escape this control, in part because not all self-antigens are expressed at the primary site of lymphocyte development or the affinity TCR-MHCII-peptide is too low [117,118]. In addition, peripheral tolerance mechanisms exist, which renders these lymphocytes tolerant, when they first encounter their cognate self-antigen outside the thymus.

1.3.2. Peripheral tolerance

Cells escaping central tolerance are kept under control by mechanisms of peripheral tolerance. Mature T cells that recognize self-antigens in peripheral tissues become incapable of subsequently respond to these antigens. The mechanisms of peripheral tolerance are: anergy, deletion, ignorance and Tregs. Anergy is characterized by the inability of a CD4 T cell to respond to stimulation. This state is induced when T cells are improperly activated for example without costimulation or with CTLA-4, which binds to B7 [119,120]. Anergy can be broken by addition of IL-2 *in vitro*. Immature DCs, which express a low level of costimulator molecules induce anergy [37]. Deletion is another mechanism by which CD4 T cells repeatedly activated by a persistent antigen will die by apoptosis in a process called activation-induced cell death [121].

In many cases, T cells simply ignore antigens present only within specialized organs. These T cells, even if of only low affinity for the antigen in question, could provoke autoimmunity if sufficient help is provided, for example, through localized production of IL-2 or through provision of cross-reactive help [122]. Another efficient control is active suppression mediated by Tregs [123,124].

1.3.2.1. T regulatory cells

T cells with suppressive capacity were described first by Gershon in the early 1970s and were called suppressive cells [125,126]. Several Treg subsets have been described and are involved in peripheral tolerance keeping the immune system under control. But they can roughly be divided into natural or inducible Tregs (iTregs). The natural Tregs cells or $CD4^+CD25^+$ T cells, which are characterized by high level of CD25 and FOXP3 expression, originate in the thymus but, importantly, they can also be generated in the periphery. The inducible Tregs including Tr1 and Th3 cells can be generated out of naïve cells and suppress target T cells (or responder cells) in a contact-independent manner by secretion of IL-10 [127] and TGF-β [58,128]. The Th3 cells are responsible for antigen tolerance induced, when the antigen is fed.

Oral tolerance discovered by Wells in 1911 refers to the oral administration of protein antigens, which induces a state of systemic non-responsiveness specific for the fed antigen. This method of inducing immune non-responsiveness has been applied to the prevention and treatment of experimental animal models of experimental autoimmune encephalomyelitis

(EAE) [129,130], rheumatoid arthritis [131], insulin dependent diabetes mellitus [132,133], transplantation [134,135] and food allergy [136]. Tolerance induction, in this model, is mediated by T cells. This has been shown in adoptive transfer experiments, in which T cells adoptively transferred from sensitized animal to naïve animal were preventing the disease. The Th3 cells were discovered originally using an oral tolerance model, in which SJL mice (susceptible to EAE) [137,138] were fed with myelin basic protein (MBP). MBP is a protein expressed in the central and peripheral nervous systems. It is recognized by autoreactive T cells, which destroy myelinated neurons leading to MS or EAE, MBP–specific TCR T cells are found in patients with MS. Tolerance induction was characterized by generation of antigen-specific cells producing high amount of TGF-β and lower amount of IL-4 and IL-10. Furthermore Th3 cells injected in mice at the time of immunization with MBP were protective against EAE development [139,140].

Tr1 cells were generated upon repeated stimulations of naïve T cells with OVA and IL-10. These cells produce high amount of IL-10 with or without TGF-β. Tr1 cells proliferate poorly after polyclonal or Ag-specific activation *in vitro* and have suppressive capacity as demonstrated by adoptive transfer in a mice model of colitis [141]. Il-10-secreting cells play an important role in allergies and transplantation [142,143]. Non-allergics have a higher number and frequency of IL-10-secreting cells, which keep the immune reaction under control when exposed to the allergen compared to atopic patients [144].

1.4. CD4$^+$CD25$^+$FOXP3$^+$ T regulatory cells

Among different types of Tregs, naturally arising CD4$^+$CD25$^+$ Treg cells are the best characterized and studied. These cells comprise 5-10% of CD4$^+$ T cells in peripheral lymphoid organs and represent a unique T cell lineage that undergoes thymic selection and migrates to the periphery [145]. Their relationship to the other subsets of regulatory cells is still not clear. Mature Treg cells can be identified by their constitutive expression of CD25, CTLA-4, PD-1, CD103, human leukocyte antigen-DR (HLA-DR), transferring receptor (CD71) and glucocorticoid-induced tumor necrosis factor receptor family-related receptor (GITR) [146-154].

The CD4$^+$CD25$^+$ Treg cells have been identified by Sakaguchi in 1995 in experiments consisting in the depletion of CD4$^+$CD25$^+$ T cells in adult mice, which resulted in the development of various autoimmune conditions (thyroiditis, gastritis, insulitis, sialoadenitis, adrenalitis, oophoritis, glomerulonephritis, and polyarthritis). Reconstitution of CD4$^+$CD25$^+$

cells within a limited period after depletion prevented these autoimmune developments in a dose-dependent fashion, whereas the reconstitution several days later was far less efficient for the prevention of the autoimmune disease [155]. As expected from these experiments, Treg cells play a major role in keeping the immune system under control and dysfunction have been found in many diseases ranging from autoimmunity, cancer to allergy and asthma. In autoimmune disease Treg cells are deficient [147,156-169] and restoration of $CD4^+CD25^+$ Treg could protect against the development of the disease [170]. In allergies harmless antigens are recognized by T cells and trigger an immune response. Several studies have demonstrated a role for the $CD4^+CD25^+$ Tregs in regulating allergic disease. The T cells from healthy non-allergic subjects do not proliferate when in contact with cows'milk antigen. However, depletion of $CD4^+CD25^+$ T cells resulted in T cell proliferation, suggesting that Tregs cells normally suppress the responses to dietary antigens [171]. $CD4^+CD25^+$ are generated in the periphery of the transplant during organ transplantation [172] or GVHD [173,174] and can stop the rejection reaction induced by the foreign antigen. Therefore Tregs are promising targets to diminish the immune reaction induced in allergies, transplantation and autoimmune diseases. One approach to use Treg as therapeutic tool is to expand them *in vitro*, thereafter inject expanded cells back to the host [175,176].

Although that a lack of suppression leads to deleterious immune response against harmless antigens, too much suppression obviously favors tumor growth and chronic infection. A higher frequency of Treg cells in peripheral blood was reported in patients with various cancer including breast cancer [177], colorectal cancer [178,179], oesophageal cancer [179], gastric cancer [179], hepatocellular carcinoma [180], leukaemia [181], lung cancer [182], lymphoma, melanoma [183,184], ovarian cancer [185] and pancreatic cancer compared to healthy individual. It has been shown that regulatory cells are recruited or generated in the periphery of the tumor in a TGF-β-mediated manner[186]. Tregs obviously actively suppress the immune response to the tumors and depletion of $CD4^+CD25^+$ T cells in mice, by *in vivo* injection of a depleting anti-CD25 antibody (PC61) resulted in suppression of growth of the tumor [157,187,188].

Tregs suppress proliferation and cytokine production from responder cell in a contact-dependent mechanism, since suppression does not happen when a cytokine-permeable membrane separates the cells. Importantly, the presence of APCs is not required, as suppression occurs in APC-free cultures. Suppression requires activation of suppressor T cells by TCR ligands or antibodies to CD3 [145]. Interestingly, the cells are activated in an antigen-specific way, whereas suppression occurs in an antigen-non-specific-manner [171]. Although the precise molecules involved in suppression are still unidentified a role has been suggested for CTLA-4, GITR and membrane-bound TGF-β [189-191]. However, the role of suppressive cytokines (IL-10 and TGF-β) is still unclear in the suppression mediated by CD4$^+$CD25$^+$ Tregs and has been challenged by the following studies: the addition of neutralizing antibodies that are specific for IL-10 or TGF-β does not reverse suppression, and CD25$^+$ T cells from *Il10* −/− mice are fully competent suppressors *in vitro* [192]. Furthermore, CD4$^+$ T cells from transgenic mice that express a dominant-negative form of the TGF-β receptor (TGFβRII) that cannot respond to TGF-β-derived signals [193] were fully suppressible. Finally, CD25$^+$ T cells isolated from young TGF-β -/- mice [194] are fully competent suppressors when mixed with CD25$^-$ T cells from wild-type mice. Thus, the potential role of TGF-β in CD25$^+$ T-cell-mediated suppression remains controversial. In addition Treg might use cytolytic activity against autologous CD4$^+$ and CD8$^+$ T cells, CD14$^+$ monocytes, and dendritic cells in order to regulate suppression. Inducible Treg express granzyme B and activated CD4$^+$CD25$^+$ Tregs express granzyme A and small amounts of granzyme B. Both subtypes displayed perforin-dependent cytotoxicity [195].

In vitro studies showed that CD25$^+$ suppressor T cells are anergic. They do not proliferate in culture, when stimulated with antibodies to CD3 or antigens unless supplemented with high doses of IL-2.

1.4.1. Tregs of thymic origin

A large part of CD4$^+$CD25$^+$ Treg cells originates from the thymus, since thymectomy before day 3 post-birth decreases the number of CD4$^+$CD25$^+$ and leads to autoimmunity [156]. Other evidences about the thymus origin of Treg are coming from their TCR repertoire. Hsieh et al determined that the TCR repertoire of thymic Treg cells was diverse and was more similar to that of peripheral Treg cells than that of nonregulatory T cells. The finding indicates that thymic Treg cells make a substantial contribution to the peripheral Treg cell population [196].

The mechanisms involved in thymic generation of CD4$^+$CD25$^+$ Tregs cells are poorly understood but it has been shown that thymocytes have already regulatory properties. A subset of CD25$^+$ cells in the CD4 single positive thmyocyte compartments with suppressive capacity was discovered. After characterization in adoptive transfer models and in *in vitro* suppression assays in mice and human, it showed that functional Treg are generated in the thymus [157,197,198]. The generation of CD4$^+$CD25$^+$ Treg cells requires MHC class II expression in the thymus, as does the generation of conventional CD4$^+$ T cells. However conventional CD4$^+$ T cells require low-affinity peptide-MHC class II interactions for positive selection, whereas Treg cells apparently require high-affinity peptide-MHC class II interactions with agonist peptides that otherwise induce negative selection [199].

The expression of self-antigens in the thymus is important in efficient *de novo* generation of $CD4^+CD25^+$ thymocytes as shown by studies that high-affinity interactions with agonist ligands expressed in radioresistant tissue [200,201] and specifically in thymic epithelial cells [202], are necessary for generation of $CD4^+CD25^+$ suggesting that in fact $CD4^+CD25^+$ escape negative selection. This process requires CD28-dependent costimulation of developing thymocytes and particularly the Lck-binding motif in the CD28 cytosolic tail initiate the Treg cell differentiation program in developing thymocytes [203].

A feature shared by Treg cells and T cells with autoimmune potential is the ability to recognize self-antigens. Treg cell recognition of self antigens was initially suggested after observations indicating that the presence of a particular organ was important for the maintenance of Treg cell-mediated tolerance to that organ [204] and nonregulatory T cells transduced with Treg cell-derived TCRs rapidly expand their populations *in vivo* and induce wasting disease in lymphopenic hosts [205].

To address that issue, Hsieh examined the naturally arising polyclonal TCR repertoire in normal thymic and peripheral regulatory and nonregulatory T cells. This 'normal' set of TCRs was then compared to the TCR repertoires found in TCRs expressed by autoreactive T cells in Foxp3-/- mice, as the spontaneous autoimmunity in these mice results from their lack of Treg cells [206]. No defect in negative selection was found in these mice, therefore the lack of a functional Foxp3 gene might allow autoreactive T cells, normally present in peripheral nonregulatory and regulatory T cell populations, to 'realize their pathogenic potential'. In agreement with that idea, activated but not naive T cells in Foxp3-/- mice often used TCRs found in the Treg cell TCR repertoire of normal mice. Thus, T cells expressing these self-reactive TCRs are not eliminated but instead are likely to contribute to pathology associated with Foxp3 deficiency. Suggesting that many autoimmune T cells in the normal nonregulatory T cell population may share the TCR specificity of their naturally arising Treg cell chaperones [196,205,207].

The maturation process of Tregs in the thymus may occur in the Hassall corpuscles, which express thymic stromal lymphopoietin (TSLP). Human TSLP activates thymic CD11c-positive dendritic cells to express high levels of CD80 and CD86. These TSLP-conditioned dendritic cells are then able to induce the proliferation and differentiation of $CD4^+CD8^-CD25^-$ thymic T cells into $CD4^+CD25^+FOXP3^+$ Treg. This induction depends on peptide-MHC-II interactions, and the presence of CD80 and CD86, as well as IL-2. $CD25^+CTLA4^+$

regulatory T cells associate in the thymic medulla with activated or mature DCs and TSLP-expressing Hassall's corpuscles, suggesting that Hassall's corpuscles have a critical role in DC-mediated secondary positive selection of medium-to-high affinity self-reactive T cells, leading to the generation of $CD4^+CD25^+$ Treg within the thymus [208].

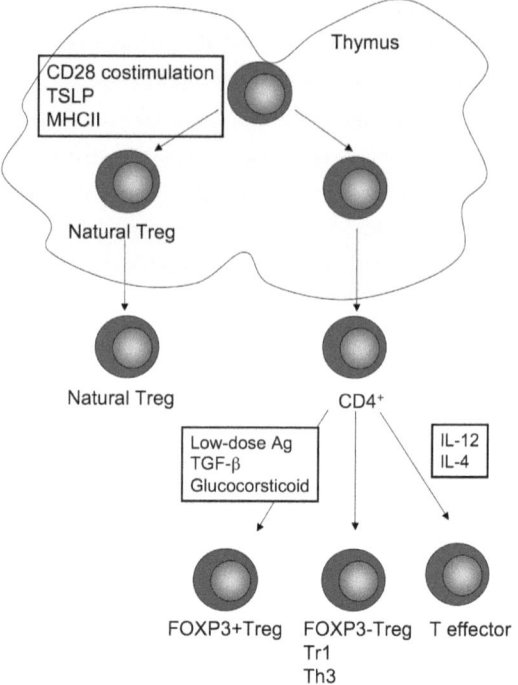

Figure 4: Generation of Tregs in the thymus and periphery. Tregs have been described to originate both in the thymus and in the periphery. The Tregs originating from the thymus are called natural Tregs whereas the Tregs from the periphery are called adaptive or inducible. The inducible Tregs are divided into different subsets, according to their FOXP3 expression or mechanisms of suppression. The Tr1 mediate suppression by IL-10 secretion and the Th3 by TGF-β.

1.4.2. Peripheral generation of CD4$^+$CD25$^+$FOXP3$^+$ Tregs

Intrathymic generation of Treg is not the only process of Treg development, importantly their generation by the conversion of CD4$^+$CD25$^-$ into CD4$^+$CD25$^+$ with suppressive capacity has been demonstrated in the periphery *in vivo* under natural conditions, allowing efficient generation of Tregs with specificity to antigens, which are not present in the thymus. The potential of CD4$^+$CD25$^-$ to become CD4$^+$CD25$^+$ was analyzed by Liang et al. in mice. CD4$^+$CD25$^-$ T cells expressing a special marker were sorted and transferred into congenic mice. After six weeks 5-12 % of transferred CD4$^+$CD25$^-$ converted to CD4$^+$CD25$^+$.

Converted CD4⁺CD25⁺ cells acquired Treg properties and phenotype, since they failed to proliferate after stimulation and could suppress proliferation of responder cells *in vitro*, and importantly also expressed high levels of Foxp3 mRNA. In addition, CD4⁺CD25⁻ cells transferred into thymectomized congenic mice converted to CD4⁺CD25⁺ Treg cells, demonstrating that the thymus is not required for peripheral generation of Tregs. Costimulation, however, was necessary since CD4⁺CD25⁻ cells transferred into B7-/- mice failed to convert into CD4⁺CD25⁺ cells that exhibit the regulatory phenotype. These results indicate that CD4⁺CD25⁻ cells convert into CD4⁺CD25⁺ regulatory T cells spontaneously *in vivo* and suggest that this conversion process could contribute significantly to the maintenance of the peripheral CD4⁺CD25⁺ regulatory T cell population. [209,210].

1.4.3. Cytokines involved in the generation and maintenance of Tregs

Cytokines are not only necessary for the function of Tregs but also for their generation and maintenance. IL-2 and IL-15 are mandatory and are their principal growth factors. However, IL-7, which is required for the development, homeostatic proliferation and maintenance of T cells [211-214], does not act on Tregs [215,216]. The IL-7Rα (CD127) is highly expressed by naïve T cells and thymocytes but is downregulated on Tregs and can be used in sorting strategies to isolate very pure Tregs population CD4⁺CD25⁺CD127⁻ [217,218]. TGF-β in addition to its many inhibitory effects on T effector cells has been shown to be important in the generation and homeostasis of Tregs.

1.4.3.1. IL-2

The constitutive high expression of the IL-2R α-chain suggests an important role for IL-2 in Treg generation and turn-over, accordingly patients receiving Il-2 therapy expand their Treg compartment [219]. The analysis of IL-2 and CD25 deficiency in mice indicated that generation of functional Treg in thymus was independent on IL-2 signaling, but that IL-2 was essential for the survival of mature CD4⁺CD25⁺ Treg in the periphery [220,221] and may be essential for their function [222,223]. Furthermore, lymphopenia seems to induce homeostatic growth of Tregs cells [224]. IL-2 triggers the JAK-STAT-signaling cascade and directly modulates FOXP3 expression. IL-2R signaling is primarily mediated through activation of JAK1 and JAK3 with subsequent phosphorylation and activation of STAT3 and STAT5 [225]. In vitro, IL-2 selectively upregulated the expression of FOXP3 in purified CD4⁺CD25⁺ T cells but not in

CD4$^+$CD25$^-$ cells. This regulation involved the binding of STAT3 and STAT5 proteins to a highly conserved STAT-binding site located in the first intron of the FOXP3 gene [226-228]. Therefore immunosuppressive drugs targeting IL-2 signalling may influence Treg turnover.

1.4.3.2. TGF-β

TGF-β has not only been proposed to be an effector cytokine secreted by CD4$^+$CD25$^+$ mediating suppression. It has also been shown to play an important role in the induction of CD4$^+$CD25$^+$ Treg out of CD4$^+$CD25$^-$ T cells *in vitro* and *in vivo* [229]. The role of TGF-β was first discovered in mice [210] and later in human [64,230]. TGF-β induces FOXP3 expression in TCR-stimulated T cells, as well as surface expression of CD25, HLA-DR, GITR, CD103 and intracellular CTLA-4 [230]. The generated cells are not only unresponsive to TCR stimulation, but produce also TGF-β and IL-10, however they do not produce Th1 nor Th2 cytokines. They are potent suppressors of proliferation and cytokine production *in vitro*. Tregs induced *in vitro* by TGF-β have been demonstrated to be suppressive *in vivo* in an OVA peptide transgenic adoptive transfer model as well as in a murine asthma model, in which TGF-β-induced Treg transferred to mice protected against the house dust mite-induced allergic pathogenesis in the lungs [210]. Although it is well accepted that TGF-β can induce CD4$^+$CD25$^+$FOXP3$^+$ Tregs *in vitro*, its role *in vivo* is still controversial. TGF-β-treated mice increase the pool of Tregs, in several experimental systems, it is not always distinguihised between truly *de novo* generation and proliferation of preexisting Tregs. TGF-β is essential in expanding Tregs as shown by a transient pulse of TGF-β in the islets of the pancreas during the priming phase of diabetes. The frequency of CD4$^+$CD25$^+$FOXP3$^+$ Tregs dramatically increased due to *in situ* expansion of Tregs [231]. TGF-β1-/- mice develop severe autoimmunity. In these mice the Tregs develop normally in the thymus, but were found in a significantly reduced number in the periphery. The Foxp3 expression in the Treg is lower and cells are less suppressive. Further indicating that TGF-β signaling is essential in maintaining Treg *in vivo* [232]. In addition transgenic mice overexpressing TGF-β under the control of the CD2 promoter show an increased frequency of Tregs in the periphery. TGF-β was also shown to enhance the conversion rate of CD4$^+$CD25$^-$ T cells to Treg when cells are stimulated with subimmunogenic peptide [233]

1.4.4. FOXP3

FOXP3 was identified by positional cloning on the X-chromosome as the gene mutated in the scurfy mouse, characterized by wasting, exfoliative dermatitis, lymphadenopathy, hepatosplenomegaly, and the presence of autoantibodies. The mouse carrying the scurfy mutation leads to death at approximately 3 weeks of age from a massive lymphoproliferative disease, with peripheral lymphocyte levels up to 20-fold greater than normal mice only male are affected. The autoimmune disease is prevented by neonatal adoptive transfer of Tregs from wild-type mice to scurfy mice [234]. In addition, the mice display anemia [235]. The autoimmune-like disease in affected animals resembles knock-out mice for *ctla-4* or *tgf-b1* genes [236,237].

In human, mutations in FOXP3 gene are responsible for the immune dysregulation polyendocrinopathy enteropathy, X-linked syndrome (IPEX; also known as X-linked autoimmunity and allergic dysregulation syndrome, XLAAD). Patients with IPEX syndrome suffer from a neonatal onset of insulin-dependent diabetes, infections, enteropathy, thrombocytopenia and anemia, endocrinopathy, eczema and cachexia. Massive T-cell infiltration into the skin and gastrointestinal tract is also observed, as well as high serum levels of autoantibodies, which is indicative of autoimmune disease. Affected children also suffer from allergic manifestations including severe eczema, high IgE levels, eosinophilia and food allergies. The severe immune dysregulation observed in human and mice lacking functional FOXP3, indicates its important function in regulating the immune system. In fact, mice lacking FOXP3 (FOXP3-/-) also lack $CD4^+CD25^+$ Treg.

Naïve $CD4^+CD25^-$ T cells, retrovirally transfected with FOXP3 were acquiring a T regulatory phenotype. They were hyporesponsive to TCR stimulation anergic and were able to suppress proliferation of other cells. Foxp3-infected T cells could suppress *in vivo* the inflammation and the autoimmune disease in a model of IBD and autoimmune gastritis that can be induced in severe combined immunodeficiency (SCID) mice by the transfer of $CD25^+CD45RBhighCD4^+$ T cells from normal BALB/c mice and prevented by cotransfer of $CD25^+CD4^+$ TR cells [238,239]. The Foxp3-transduced cells inhibited weight loss, diarrhea, and histological development of colitis and gastritis induced by the transfer of $CD25^-CD45RBhighCD4^+$ cells as effectively as naturally occurring $CD25^+CD4^+$ Treg cells [240]. Underlying the decisive role of FOXP3 in T regulatory cells development or/and function in human and mice [241].

These results are still controversial, particularly in human cells in which two isoforms of FOXP3 are found. Ectopic expression of the two FOXP3 isoforms in $CD4^+$ cells resulted in induction of hyporesponsiveness and suppression of IL-2 production, but the cells were only weak suppressors. These data indicate that in humans, overexpression of FOXP3 alone or together with FOXP3delta2 is not an effective method to generate potent suppressor T cells *in vitro*. And suggest that factors in addition to FOXP3 are required during the process of activation and/or differentiation for the development of Tregs [242].

Using mice harboring a GFP-Foxp3 fusion protein-reporter knockin allele it has been shown that FOXP3 expression is restricted to a subset of $\alpha\beta$ T cells, which are $CD25^+$ but can also be CD25-. Importantly FOXP3 expression correlates with regulatory and suppressive function [243-245]. FOXP3 is therefore seen as a lineage factor for commitment into Tregs. FOXP3 belongs to a large family of functionally diverse transcription factors based on its winged helix-forkhead DNA-binding domain (forkhead box(Fox)). These proteins have been classified into subfamilies (indicated by the letter after "FOX") based on phylogenetic analysis of homology in the forkhead domain only, and each has been assigned a unique number (at the end of the name) [246]. In addition to the C-terminal forkhead domain, FOXP3 also contains a Cys2His2 zinc finger domain and a coiled-coil-leucine zipper motif. Homology among full-length human, mouse and rat FOXP3 is very high, suggesting a highly conserved function. Members of the Fox family are both transcriptional activators and transcriptional repressors. The ability of Foxp3 to act as a transcriptional repressor required the presence of the FKH domain of Foxp3. At present there is very little understanding of the function of FOXP3 at the molecular level. FOXP3 binds DNA, localizes to the nucleus and can act as a transcriptional repressor. In the Jurkat T cell leukemia cell line, it inhibits transcription mediated by the nuclear factor of activated T cells (NFAT) transcription factors, requiring the forkhead domain for both nuclear localization and DNA binding. FOXP3 can also form a complex with NFAT while competing with AP-1. And therefore an activating complex NFAT-AP1 is replaced by a repressive complex NFAT-FOXP3 [247]. In addition mice expressing a Foxp3 transgene were unable to produce IL-2, IL-4 or IFN-y following TCR-mediated stimulation in vitro and showed a severely reduced ability to express cytokines in vivo following immunization [248,249].

The Foxp3 transgenic mice also provided a model system for examining the in vivo consequences of Foxp3 expression. When bred into an otherwise wild-type background, resulted in a reduction in peripheral $CD4^+$ and $CD8^+$ T cell numbers [234].

However thymic cellularity was unaffected, as was positive and negative selection. Thus, levels of Foxp3 determined the number of peripheral T cells, while having little effect on the number and differentiation of thymocytes [250].

1.5. Concluding remarks and aim of the study

T cells with suppressive capacity have been identified in the 1970s. Due to difficulties in isolating and characterizing the cells enthusiasm dampened down and scepticism was growing. In 1995, Sakaguchi identified $CD4^+CD25^+$ T cell as potent regulators of the immune response, the field was thus, revitalized. Treg dysfunctions are found in many diseases bringing a new understanding of many pathogenesis. Therapies targeting Tregs are therefore very promising. CD25, the IL-2 receptor α chain, is not a reliable marker, since it is upregulated during activation of $CD4^+$ T cells. Therefore the discovery of FOXP3, as a transcription factor expressed selectively in Tregs, was a great breakthrough in tolerance immunology. Although it is an intracellular protein, development of specific antibodies and transgenic animals allows analysis at the single cell level. A better knowledge of its gene expression regulation will give a better understanding of the Treg turnover.

The aim of this thesis was to gain insight into the mechanisms involved in Treg generation. For this purpose we looked at the FOXP3 gene regulation, particularly we analyzed its promoter and identified molecular pathways involved in the induction and repression of FOXP3 expression.

2. Results

2.1. Molecular mechanisms underlying FOXP3 induction in human T cells[1]

Pierre-Yves Mantel,* Nadia Ouaked,* Beate Rückert,* Christian Karagiannidis,* Roland Welz,* Kurt Blaser,* Carsten B. Schmidt-Weber*

Keywords: Human, T Cells, transcription factors, gene regulation

Corresponding author: Carsten B. Schmidt-Weber, Swiss Institute of Allergy and Asthma Research (SIAF), Obere Str. 22, CH-7270 Davos, Switzerland;
e-mail: Carsten.schmidt-weber@siaf.unizh.ch;
Tel.: ++41 81 410 08 53, FAX: ++41 81 410 08 40

Publishsed in: The Journal of Immunology, 2006, 176: 3593-3602

---

[1] This work was supported by the Swiss National Foundation Grants Nr: 31-65436 and 3100A0-100164, the Ehmann Foundation Chur, the Ernst Goehner Foundation Zug, the Saurer Foundation Zurich and the Swiss Life Zurich.

* Swiss Institute of Allergy and Asthma Research (SIAF), Obere Str. 22, CH-7270 Davos

Abstract

FOXP3 is playing an essential role for T regulatory cells (Tregs) and is involved in the molecular mechanisms controlling immune tolerance. Although the biological relevance of this transcription factor is well documented, the pathways responsible for its induction are still unclear. The current study reveals structure and function of the human FOXP3 promoter, revealing essential molecular mechanisms of its induction.

The FOXP3 promoter was defined by RACE, cloned and functionally analyzed using reporter-gene constructs in primary human T-cells.

The analysis revealed the basal, T-cell-specific promoter with a TATA and CAAT-box 6000bp upstream the translation start site. The basal promoter contains six NFAT and AP-1 binding sites, which are positively regulating the transactivation of the FOXP3 promoter after triggering of the TCR. The chromatin region containing the FOXP3 promoter was bound by NFATc2 under these conditions. Furthermore, FOXP3 expression was observed following TCR engagement. Both promoter activity, mRNA and protein expression of T-cells were suppressed by addition of cyclosporin A (CsA). Taken together, this study reveals the structure of the human FOXP3 promoter and provides new insights in mechanisms of addressing T_{reg} inducing signals useful for promoting immune tolerance. Furthermore the study identifies essential, positive regulators of the FOXP3 gene and highlights CsA as an inhibitor of FOXP3 expression contrasting other immunosuppressants such as steroids or rapamycin.

Introduction.

T cells play a key role in adaptive immunity and enable the immune system to develop specific immune responses. T cell activation is tightly regulated allowing responses against pathogens, while maintaining tolerance of harmless antigens. Disequilibrated immune tolerance causes autoimmune disease or allergy. Thus immune tolerance is an important mechanism that allows to distinguish between self and non-self [251,252]. Regulatory T cells (Tregs) are critical regulators of immune tolerance and their suppressive control of effector T cells was observed in experimental systems [253] and human [144,254]. Tregs are defined by their function, and express the transcription factor FOXP3 and/or suppressive cytokines (IL-10, TGF-β) as well as CTLA-4 and/or CD25 [146,155,192,255,256]. FOXP3 is a transcription factor, belonging to the forkhead family [257] and it has been shown that FOXP3, overexpressed in Jurkat cells, can act as a repressor of transcription of the IL-2 promoter by competing with the binding of NFAT [258] or by directly interacting with NFAT or NFκB [259]. The CD25$^+$ Tregs express constitutively high amount of FOXP3 and represent about 5-10% of the total CD4$^+$ population. Despite the great relevance of these cells in immunology and clinical issues, the origin and pathways of Treg induction are still unclear. Interestingly it could be demonstrated that ectotrophic expression of FOXP3 in T cells was sufficient to restore autoimmune symptoms of mice depleted of CD25$^+$ T cells [220,240]. In fact genetic defects of the human ortholog causes the IPEX syndrome (immune dysregulation polyendocrinopathy enteropathy, X-linked) [257]. Patients with IPEX syndrome suffer from a neonatal onset of insulin-dependent diabetes, infections, enteropathy, thrombocytopenia and anemia, endocrinopathy, eczema and cachexia [260] and transgenic mice lacking FOXP3 are developing a severe autoimmune disease [248,250,261].

These evidences indicate that FOXP3 is a gene, which is involved in the generation or maintenance of regulatory T cell phenotypes, which is essential for maintaining immune tolerance. Interestingly it has been shown that its expression can also be induced in the CD4$^+$CD25$^-$ population by activation [63], corticosteroids [262], estrogen [263] and TGF-β [264,265], suggesting that FOXP3 can be induced in peripheral T cells, which may become crucial for therapeutic interventions. We therefore investigated the FOXP3 promoter to systematically reveal signals inducing FOXP3 expression.

The 5'-flanking region of the human FOXP3 gene was cloned and the promoter activity was characterized in primary $CD4^+$ T cells. The data demonstrate that the proximal promoter is localized in the region between $-511/+176$ bp upstream the 5'-non-coding region and contains several common features of basal promoter such as a GC and a TATA box. Our results demonstrate that the promoter is inducible by activation in a NFAT-AP-1 dependent manner, which is inhibited by CsA.

Materials and Methods

Localization of the human FOXP3 promoter by RACE
The cDNA from CD4$^+$ T-cells was amplified with the anchor primer and two nested antisense primers: RACE FOXP3 +987, RACE FOXP3 +521 (Table I) designed from the FOXP3 cDNA sequence. The PCR products were purified and cloned into pCR2.1 vector (Invitrogen, Basel, Switzerland) for sequencing of the 5'cDNA ends.

Cloning of the FOXP3 promoter, construction of deletion and mutant constructs
The human FOXP3 promoter containing -1657 bp from TSS was amplified by PCR using FOXP3 promoter sequence specific primers from position -1657 to +176. The genomic DNA extracted from CD4$^+$ T cells of a healthy donor was used as a template. The FOXP3 promoter amplicon was cloned into the pGL3 basic vector (Promega Biotech Inc., Madison, WI, USA) to generate the pGL3 FOXP3 – 1611/+176. Series of deletion constructs were generated. The PCR products were subcloned in the pGL3 basic vector. Site-directed mutagenesis in the FOXP3 promoter region were introduced using the QuickChange kit (Stratagene, Amsterdam, The Netherlands), according to the manufacturer's instructions. The constructs were generated by using pGL3 –511, -348, –307 or –211 as template. Primers which were utilized to generate the individual constructs are listed in Table I.

Bioinformatics
Genomic sequences spanning the 5'-UTR of the FOXP3 gene was analyzed using the alignment software m-Vista: http://www-gsd.lbl.gov/vista/VistaInput [266], allowing to identify conserved regions. Transcritption factor binding sites were identified using TESS (http://www.cbil.upenn.edu/cgi-bin/tess/tess33) and Genomatix (http://www.genomatix.de) program, which uses matrices of the Transfac database.

Isolation of CD4$^+$ T cells
CD4$^+$ T cells were isolated from blood of healthy volunteers using the anti-CD4 magnetic beads (Dynal) as previously described [267]. The purity of CD4$^+$ T cells was initially tested by FACS and was ≥ 95%.

Flow cytometry

For analysis of FOXP3 expression at the single-cell level, cells were first stained with the monoclonal antibody CD25 (Beckman & Coulter, Switzerland), after fixation and permeabilization, cells were incubated with phycoerythrin-conjugated monoclonal antibody PCH101 (anti-human FOXP3; eBioscience) based on the manufacturer's recommendations and subjected to FACS (EPICS XL-MCL, Beckman& Coulter).

Transfections and reporter gene assays

T cells were rested in serum-free AIM-V medium (Life Technologies, Basel, Switzerland) overnight. An amount of 3.5 µg of the FOXP3 promoter Luciferase reporter vector and 0.5 µg phRL-TK was added to 3 x10^6 CD4$^+$ T cells resuspended in 100 µL of Nucleofector^{TM} solution (Amaxa Biosystems, Cologne, Germany) and electroporated using the U-15 program of the Nucleofector^{TM}. After a 24 hour culture in serum-free conditions and stimuli as indicated in the figures, luciferase activity was measured, by the dual luciferase assay system (Promega Biotech Inc., Madison, WI, USA) according to the manufacturer's instructions. Data were normalized by the activity of renilla luciferase. Hela, CHO and Jurkat were transfected using lipofectamine 2000 (Invitrogen) according to the manufacturer's protocol.

Quantitative real-time PCR

The PCR primers and probes detecting FOXP3 were designed based on the sequences reported in GenBank with the Primer Express software version 1.2 (Applied Biosystems) as follows: EF-1α forward primer and reverse primer as described [268], FOXP3 forward primer A 5` GAA ACAG CAC ATT CCC AGA GTT C 3`, FOXP3 reverse primer A 5` ATG GCC CAG CGG ATG AG 3`. The prepared cDNAs were amplified using SYBR$^®$-PCR mastermix (Applied Biosystems) according to the recommendations of the manufacturer in an ABI PRISM 7000 Sequence Detection System (Applied Biosystems). Relative quantification and calculation of the range of confidence was performed using the comparative ΔΔCT method as described [269]. All amplifications were carried out in triplicates.

Western blotting

For FOXP3 analysis on the protein level, 1x10^6 cells CD4$^+$CD25$^-$ were lysed and loaded next to a protein-mass ladder (Magicmark, Invitrogen) on a NuPAGE 4-12% bis-tris gel (Invitrogen). The proteins were electroblotted onto a PVDF membrane (Amersham Life Science, Dübendorf, Switzerland). After blocking the membranes were incubated with an 1:200 dilution of goat anti-FOXP3 in blocking buffer (Abcam, Hamburg, Germany) overnight at 4°C. The blots were developed using an anti-goat HRP labeled mab (Amersham Biosciences) and visualized with a LAS 1000 camera (Fuji, Urdorf, Switzerland). To confirm sample loading and transfer, membranes were incubated in stripping buffer and re-blocked for 1 h, then re-probed using anti-actin (C-2, Santa Cruz).

Pull-down Assay

CD4$^+$ T cells were stimulated with PMA and ionomycin for 2 hours at 37°C. The cells were pelleted, resuspended in buffer C (20 mM HEPES (pH 7.9), 420 mM NaCl, 1.5 mM MgCl$_2$, 0.2 mM EDTA, 1 mM DTT, protease inhibitors (Sigma, Buchs, Switzerland) and 0.1% NP-40) and lysed on ice for 15 min. Insoluble material was removed by centrifugation. The supernatant was diluted 1:3 with buffer D (as buffer C, but without NaCl). The lysates were incubated with 10 µg of poly(dI-dC; Sigma) and 70 µl of streptavidin-agarose (Amersham Biosciences) carrying biotinylated oligonucleotides, for 3 hours at 4 °C. The beads were washed twice with buffer C/D (1:3) and resuspended in DTT-containing loading buffer (NuPAGE; Invitrogen), heated to 70°C for 10 min and the eluants loaded next to a protein-mass ladder (Magicmark, Invitrogen) on a NuPAGE 4-12% bis-tris gel (Invitrogen). The proteins were electroblotted onto a PVDF membrane (Amersham Life Science, Dübendorf, Switzerland) and detected using an anti-NFATc mab (Santa Crusz). The blots were developed as described above. Accumulated signals were analyzed using AIDA software (Raytest, Urdorf, Switzerland).

Electrophoretic mobility shift assay (EMSA)

EMSAs were performed as previously described [270]. Nuclear extracts of CD4$^+$ T cells were prepared as described. Briefly, the cells were treated with a hyposmotic buffer, containing 10mM KCl, 1mM DTT, 0.5 mM EDTA, 10 mM HEPES pH 7.9 (all Sigma) and a mix of protease inhibitors (Complete™; Boehringer Mannheim, Mannheim, Germany), followed by addition of NP-40 (Sigma) to a 1% final concentration. Nuclei were pelleted by a brief spin in a microcentrifuge and washed once with the buffer described above. Nuclei were lysed in 50 µl of a high-salt buffer containing 400mM NaCL, 50mM DTT, 20 mM HEPES, 0.5 mM EDTA and a mix of protease inhibitors (Complete™, Boehringer Mannheim, Mannheim, Germany). The nuclear debris of this lysate was removed by centrifugation at 4°C and the supernatant stored in a fresh tube at -70°C. Nuclear extracts were controlled for equal protein content by a protein assay as described by the manufacturer (Biorad, Hercules, CA, USA).

Nuclear extracts were incubated with annealed oligonucleotides (Table I), which correspond to the FOXP3 promoter sequences as indicated in the figures. The two strands of the oligonucleotides first labeled with ^{32}P-γ-ATP using the T4 Kinase (Life Technologies). Subsequently, the oligonucleotides were separated from free ^{32}P-γ-ATP by running the labeling mix over a chromaspin-10 column (Clontech, Palo Alto, CA). Following annealing, single stranded oligonucleotides were eliminated by gel-purification of the column eluate on a 20% polyacrylamide gel. The eluted probe was precipitated and the binding reactions for the TATA-site were carried out for 30 min at RT with 2 µg of NE in 10 mM HEPES (pH 7.9), 10% glycerol, 1 mM EDTA, 1 mM DTT, 100 mM KCl, 0,5 ug of poly(dI-dC), 1 mM PMSF and 30 000 cpm of probe. For the GC-box, 3 µg of NE were incubated as previously described in [271]). The reaction was incubated for 10 min. at room temperature and loaded on a 5% non-denaturating PAA gel. Following electrophoresis, the gel was dried, subjected to autoradiography and phosphoimaging.

Chromatin immunoprecipitation (ChIP) assay

ChIP assay was performed chromatin Immunoprecipitation (ChIP) Assay Kit following the recommandations of the supplier (Upstate Biotechnology, Lake Placid, NY, USA). For precipitation a polyclonal Ab against acetylated histone H4 was used along with an isotype matched rabbit IgG control. The PCR addressed for the FOXP3 promoter region -246 to -511 and was performed using the following primers: 5'-GTG CCC TTT ACG AGT CAT CTG-3' and 5'-GTG CCC TTT ACG AGT CAT CTG-3'. The PCR products were visualized using an ethidium bromide gel. For ChIP assay addressing the NFAT binding to the chromatin an anti-NFATc2 (4G6-G5, Santa-Cruz Biotechnology, Santa Cruz, CA, USA) was used and primer addressing the FOXP3 promoter region –1540 to -1470 to the following primer were used 5'-TTT GCA GGG TGC TGG GA-3' and 5'-GTA GAC CAG CCC CCA GGG-3' and qRT- PCR was performed.

FACS-sorting of $CD4^+CD25^+$

PBMC were isolated from Buffy coat by density gradient centrifugation over Ficoll/Hypaque. Cells were stained with PE-anti-CD25 and anti-PE magnetic beads (Miltenyi Biotec, Bergisch Gladbach, Germany) and $CD25^+$ cells were enriched using the Midi-MACS system (Miltenyi Biotec). CD25-enriched or -depleted cell populations were stained with FITC-anti-CD4 and sorted into $CD4^+CD25^-$ and $CD4^+CD25^{high}$ on a FACStar Plus (BD Biosciences).

Suppression assay

Samples in triplicate, containing 5×10^4 $CD4^+CD25^-$ and 1×10^4 of preactivated or resting $CD4^+CD25^+$T cells per well were incubated in 96 round-bottom-plates, which were previously coated with 1µg/ml antiCD3 mab or a matched isotype control. Cells were cultured for 4 days, pulsed for the last 10 h with 1 µCi [^{3}H]-thymidine (Hartmann, Braunschweig, Germany) and harvested on glass fiber filters using an automated multisample harvester (LKB, Pharmacia-Wallac, Turku, Finland). Filters were transferred in sample bags with liquid scintillation fluid and analyzed using a β-scintillation counter (Pharmacia-Wallac).Round-bottom 96-well plates were coated with 1 µg/µl anti-CD3 for 1 hour at 37 °C and subsequently washed with PBS.

Results.

Localization of the FOXP3 promoter in human CD4$^+$ T cells

To map the 5' end of the human FOXP3 gene, 5'-RACE was performed, using nested-PCR. The first primer was located in exon 11 and the second in exon 6 of the FOXP3 gene. The mRNA was isolated from CD4$^+$ T cells of a healthy donor, reverse-transcribed and used as template for 5'-RACE. Sequence analysis of 11 clones revealed that the transcription start site (TSS) is located 6211 bp upstream of the translation start site. The UTR is interrupted by an intron zero of 6011 bp. An alignment of the sequences of human, mouse and rat FOXP3 gene was performed and several conserved regions (Figure 5A) were identified including 11 exons (Figur 5A, dark blue) and some conserved non-coding sequences (Figure 5A, red, CNS). Interestingly the region preceding the UTR is highly conserved (Figure 5A and Figure 5B) and contains several transcription factor binding sites. On the basis of these sites, a putative promoter scheme was generated and tested in the following experiments (Figure 5C).

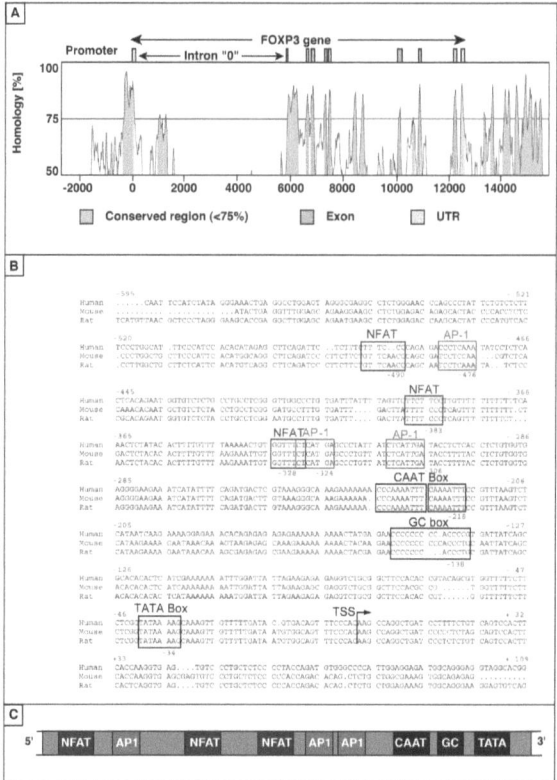

Figure 5: Human, mouse and rat alignment of the FOXP3 core promoter. (A) m-Vista alignment of Human/mouse genomic sequences (human accession number AF235097, mouse accession number AF277994). m-Vista criteria which were applied require 75 % identity for at least 100 bp length. The conserved regions are in red, the exons in dark blue and the UTR in light blue. (B) Sequence conservation of the human (top: GenBank accession number AF235097), mouse (middle: accession number AF277994) and rat (bottom: GenBank accession number NW_048035). The transcription start site (TSS) is indicated by a broken arrow. Transcription factor binding to the regions of interest are indicated (factor name above and position below), (C) Scheme of the 5'UTR region of the human FOXP3 gene, indicating the sites analyzed in this study.

Chromatin structure

Since FOXP3 is specifically expressed in T cells, we analyzed whether chromatin in the area of the putative promoter is accessible to the transcriptional machinery in T cells by chromatin co-immunoprecipitation (ChIP). Histone H4 hyperacetylation (H4ac) is a typical feature of active transcription [272], we therefore analyzed chromatin hyperacetylation of the FOXP3 promoter by comparing cells of lymphoid and non-lymphoid origin as well as T cells characterized by low or high FOXP3 expression. T cells were depleted of CD25$^+$ and intracellular FACS staining revealed that FOXP3 expression was virtually absent in the remaining cells (0.4%; figure 6A). The frequency of FOXP3$^+$ T cells increased following T cell activation predominantly in the CD25$^+$ subset (11.1 % after 72h; figure 6A). It occurred possible that activation-induced FOXP3$^+$ T cells expand from the 0.4% of CD25$^-$FOXP3$^+$ T cells, however on the basis of known T cell division kinetics (doubling maximally in 48h), the FOXP3 expression must predominantly arise from the FOXP3$^-$ T cells. We demonstrated that histone hyperacetylation is detectable in CD4$^+$ T cells, particularly in activated or FACS-sorted CD25^{+high} Treg cells, but absent in HELA and Jurkat cells. Lower levels were observed in resting CD4$^+$CD25$^-$ and CD4$^+$CD45RA$^+$ T cells (figure 6B), showing that the FOXP3 promoter region is in an open conformation and accessible to the transcription machinery in the CD4$^+$CD25$^-$ cells, and that activation might play an important role in mobilizing the chromatin structure. The acetylation levels correspond to the FOXP3 mRNA expression levels of these cells (figure 6C).

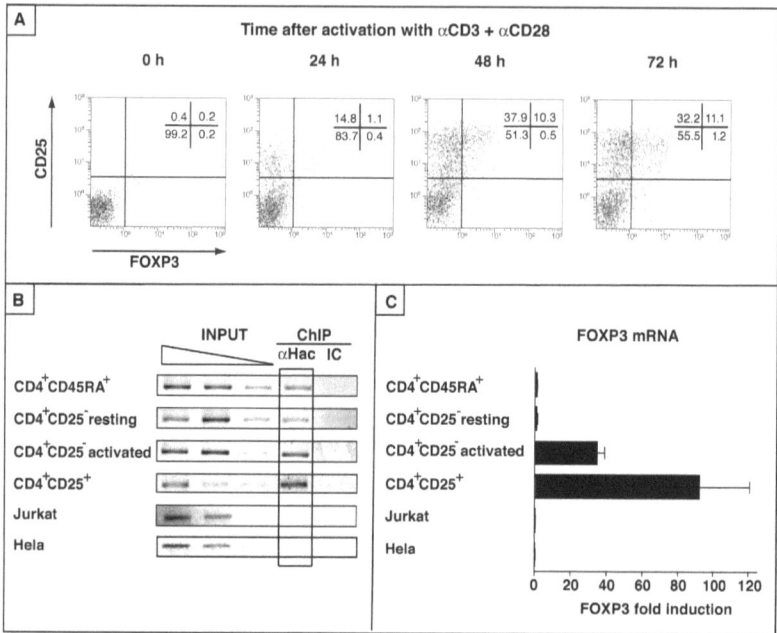

Figure 6: Chromatin configuration of FOXP3. (A) FACS staining indicating FOXP3 expression in CD4+CD25- T cells. Representative of two independent experiments. (B) The acetylation status of histone H4 in the nucleosomes associated with the FOXP3 core promoter region was assessed by ChIP assay in Jurkat, Hela and CD4$^+$CD25$^-$ (resting and activated), and cells. Cells were lysed, and proteins were cross-linked with formaldehyde and immunoprecipitated with Ab to acetylated histone H4 (anti-acetyl H4) or control Ab (rabbit IgG). Shown is the, PCR for the FOXP3 gene after reversing the cross-linking. The "input" represents PCR amplification of the total sample which was not subjected to any precipitation. Results are representative of three independent experiments. (C) expression level of FOXP3 mRNA measured by RT-PCR. Bars show the mean ± SD of three independent experiments.

The FOXP3 promoter region contains cell-specific activity

The chromatin accessible region described above was functionally investigated for transactivational activity. To identify potential regulatory elements in the 5'-flanking region of the human FOXP3 gene, a serie of promoter-luciferase (LUC) 5'-deletion constructs were generated to test whether the FOXP3 promoter fragment also reflects cell-specificity. We transfected the identical constructs into cells of lymphoid and non-lymphoid origin that do not express FOXP3. High transactivation was observed in primary CD4$^+$ T cells, whereas no activity was detected in Hela, CHO (data not shown) nor Jurkat cells independently of the promoter-fragment size (Figure 7). The longest construct was designed from position –1657 to +176 and displayed a promoter activity in CD4$^+$ T cells 3-fold higher than that of the control plasmid, pGL3 Basic (Figure 3). We designed 5'-deletions (-1210, - 511, -465, -423, -348, -307, –211 and -90) in order to identify the proximalpromoter, which we could localize in a fragment of –511 bp from transcription start site. The –511/+176 region is highly conserved between human, mice and rat (Figure 5A + 5B). A 6.8-fold increase in luciferase activity was measurable with the fragment of –511/+176 compared to pGL3 basic vector. In contrast the smaller constructs (–307, -211 or -90/+176) show lower luciferase activity. Although the construct –307/+176 shows low activity, it is essential for the activity of the –511/+176, since a deletion of –245 to + 176 region out of –511 (-511/-245 construct) shows no activity in CD4$^+$ T cells (Figure 7). Thus the construct –511/+176 showed the most prominent reporter activities, whereas larger fragments didn't show any significant increase in activity over the –511/+176 construct. These results together with the open chromatin configuration suggest, that the first 500 bp of basal FOXP3 promoter confer cell- specificity and transactivation. Having demonstrated that the promoter is active in CD4$^+$ T cells we performed site-directed mutagenesis to further characterize the promoter.

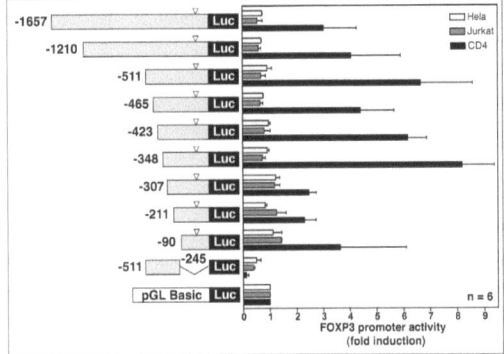

Figure 7: The putative FOXP3 promoter is tissue specific. *T cells, Jurkat and Hela cells were transfected with empty vector or vector containing the putative FOXP3 promoter region. Bars show the mean ± SD of arbitrary light units normalized for renilla luciferase of experiments performed with 6 independent donors (CD4$^+$ T cells) or 6 independent experiments (in the case of Jurkat and Hela cells; samples were measured as triplets).*

Basal transcriptional elements are located in the core promoter

To further investigate the functionality of the basal FOXP3 promoter located within the first 500 bp, we investigated binding sites characteristic for eukaryotic promoters. Putative Transcription factors binding sites were identified using TESS and GENOMATIX programs. In fact, several common features of eukaryotic core promoters such as the TATA, GC and CAAT boxes were identified. The TATA box (TATAAAA) is located –44 bp upstream of the transcription start site. This sequence is conserved between human, mouse and rat (Figure 5A). Since the TATA box is an important feature of eukaryotic promoters and is generally located -30 to -25 bp upstream the TSS, we investigated the element using site-directed mutagenesis of the fragment –211/+176 (TATAAAAG was mutated to TcTcgAAGC) and could demonstrate that the mutations, which eliminate TATA-binding sites dramatically reduce by 47.64 %; (Figure 8A) reporter activity. EMSA of the TATA box sequence of the FOXP3 promoter (TTA GAA GAG ACT CGG TAT AAA AGC AAA GTT GTT TT) bound by nuclear extracts from $CD4^+$ T cells confirmed that nuclear proteins are binding to this promoter element. Only one complex could be detected, which could be competed by pre-incubation with unlabelled oligonucleotides specific for TATA box consensus sequence (GCA GAG CAT ATA AAA TGA GGT AGG A), which abolished in a dose-dependent manner the formation of the complex (Figure 8B).

The GC box is another basic element of eukaryotic promoters, which is located 138 bp upstream the TSS. A site-specific mutation (GC Sp1 - 142) was introduced to destroy transcription factor binding site into the–307/+176 fragment and luciferase assays were carried out. The mutation in the GC box decreased transactivational activity by 42,84 % (Figure 4C). The GC-box is known to be bound by Sp transcription factor family members. Sp1 acts as a potent activator and Sp3 can act as an activator or a suppressor, possibly by competing with Sp1 for the binding. Nuclear extracts from $CD4^+$ T cell formed two specific complexes (Figure 8D), which were dose-dependently competed by the addition of specific Sp1-binding oligonucleotides at a 10 x and 100 x molar excess (ATT CGA TCG G__GG__ CGG GGC GAG C) but not by mutated Sp1 oligonucleotides (ATT CGA TCG GTT CGG GGC GAG C). The addition of an antiserum against Sp1 shifted a band on the EMSA and the remaining complex I or II migrate slightly faster, indicating that the complex becomes smaller.

Similar observations were made using an anti-Sp3 antiserum, demonstrating that Sp1 and Sp3 are binding to this sequence. Of note, the GC-box sequence can be bound also by other factors, which explains the binding of slightly faster migrating complexes upon addition of anti-SP-1 & 3 antibodies (lane 7-9; [273]). Furthermore the CAAT-box was analyzed and mutated as described for the TATA and GC box. The mutation in the CAAT also reduced the luciferase activity of the -307/+176 fragment (data not shown).

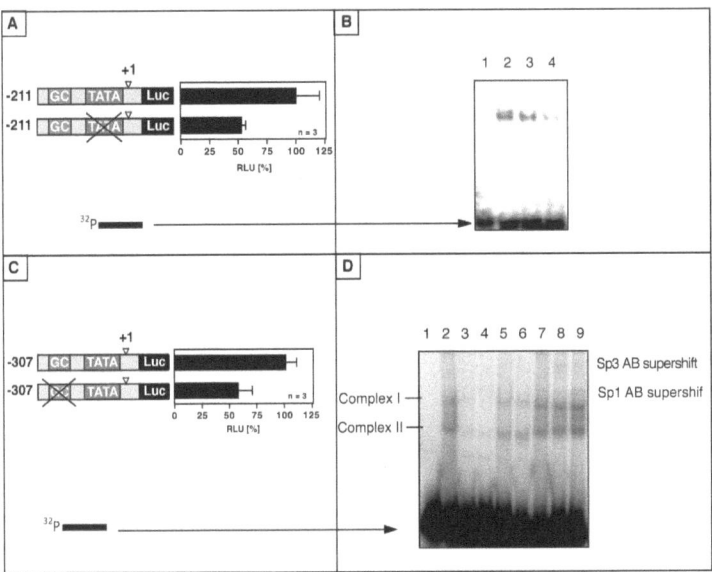

Figure 8: Basal elements of the human FOXP3 promoter. (A) The human FOXP3 contains functional TATA box and GC box. The region −211/+176, which contains the TATA box, was mutated in the pGL3 FOXP3 −211/+176. The mutated TATA box transfected into CD4$^+$ T cells and the luciferase activity was measured. The effect of mutagenesis is shown as percent relative to wild-type pGL3 FOXP3 −211/+176. Results are given as the mean ± SEM of three independent experiments in triplicate. (B) Binding of specific nuclear factors to the TATA box. EMSA of the region −60 to −14 region of the FOXP3 gene promoter is shown. The competition experiments were performed by preincubating nuclear extracts with 10- and 100-fold excess of TATA oligonucleotides (lanes 3-4). The region −307/+176 which contains the GC box was mutated in the pGL3 FOXP3 −307/+176. (C) The mutated GC box was transfected into CD4$^+$ T cells and the luciferase activity was measured. Effect of mutagenesis is shown as percent relative to wild-type pGL3 FOXP3 −307/+176. Results are shown as the mean ± SEM of three independent experiments performed in triplicate. (D) Binding of specific nuclear factors to the GC box is demonstrated by EMSA of the −124 to −173 region of the FOXP3 gene promoter. The competition experiments were performed by preincubating nuclear extracts with 10- and 100-fold excess of Sp1 oligonuclotides (lanes 3-4) or mutated Sp1 oligonucleotides (lanes 5-6). The supershift assays were performed with antiserum against Sp1 protein (Sp1 Ab; lane 7), Sp3 protein (Sp3 Ab, lane 8), Sp1 and Sp3 proteins (Sp1 + Sp3 Abs; lane 9). Supershifted bands are observed in lanes 7-9 along with an

increased mobility of the remaining bands carrying unidentified factors, since the GC-box is bound by multiple factors.

Regulation of FOXP3 expression in the $CD4^+CD25^-$ cells by activation of the T cell receptor

The experiments showed that the promoter construct was active in lymphocytes and contains basic elements like a TATA and a GC box. Since T cell activation is important for regulation of immune-relevant genes, we investigated, whether FOXP3 mRNA and FOXP3 promoter fragments respond to T cell activation. FOXP3 mRNA can be induced (18.8-fold at 24h, 30.4-fold at 48h and 11.7-fold at 72h; Figure 9A) in the $CD4^+CD25^-$ following T cell activation. Resting $CD4^+CD25^+$ were used as a control. Of note, FOXP3 expression in resting $CD4^+CD25^+$ T cells was 80-fold higher than in $CD4^+CD25^-$ cells and could just slightly be increased by activation (1.9-fold, Figure 8 B). In analogy to upregulated FOXP3 mRNA, T cell activation also induced FOXP3 reporter activity in the $CD4^+CD25^-$ cell fraction. The smaller fragments and empty vector were only slightly responsive to activation in contrast to the fragments starting from -348/+176, which were strongly induced about 80-fold compared to the empty vector or 10-fold compared to the corresponding unstimulated cells (Figure 9B).

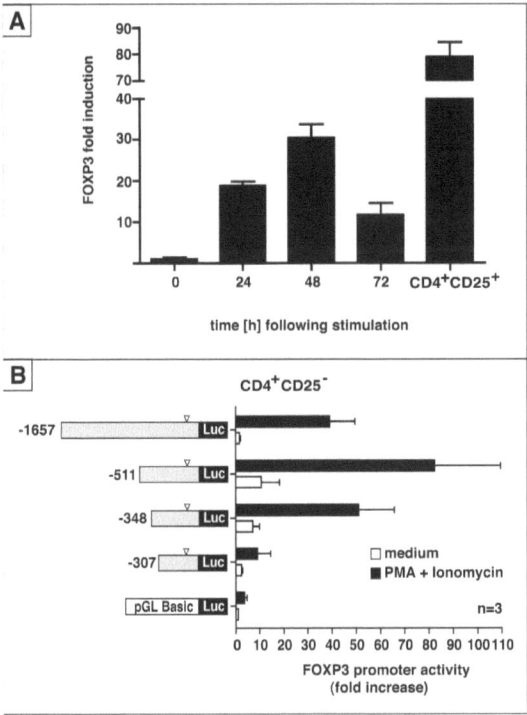

Figure 9: FOXP3 is upregulated by TCR crosslinking. (A) CD4$^+$CD25$^-$ T cells were stimulated with anti-CD3 and anti-CD28, and the FOXP3 mRNA level was measured by real-time PCR. Bars show the mean ± SD of 3 independent experiments. (B) The FOXP3 promoter can be activated by TCR cross-linking. CD4$^+$CD25$^-$ T cells were co-transfected with a renilla luciferase vector plus the luciferase vector containing the putative promoter region and were cultured in medium or in medium containing PMA and ionomycin. Results given are the mean ± SD of luciferase light units normalized for renilla luciferase of the same sample. Results are representative of 3 independent experiments.

Cyclosporin A inhibits FOXP3 expression in human CD4$^+$CD25$^-$ T cells

We identified NFAT and AP-1 transcription factors binding sites located in the region between -348 and -511, which are known to be involved in T cell activation (Figure 5B). NFAT is activated by the Ca^{++}-calcineurin pathway and blocked by the immunosuppressive drug CsA. Therefore we analyzed the effect of CsA on the induction of FOXP3 mRNA and promoter activity. CD4$^+$CD25$^-$ T cells were activated in the presence or absence of CsA and the mRNA was quantified after 24, 48 and 72 hours. The FOXP3 mRNA was potently inhibited by CsA, but not by MAPK inhibitors (Figure 10A). CsA inhibition of FOXP3 mRNA induction was maintained throughout the 72 hours time course (Figure 10B), while cell viability was maintained (data not shown). The CsA sensitivity of FOXP3 was confirmed at the protein level (Figure 10C). FOXP3 promoter fragments, which responded to activation, were potently inhibited by CsA. This indicates that the calcineurin-dependent NFAT mobilization plays a crucial role in the transactivation of the FOXP3 promoter (Figure 10D).

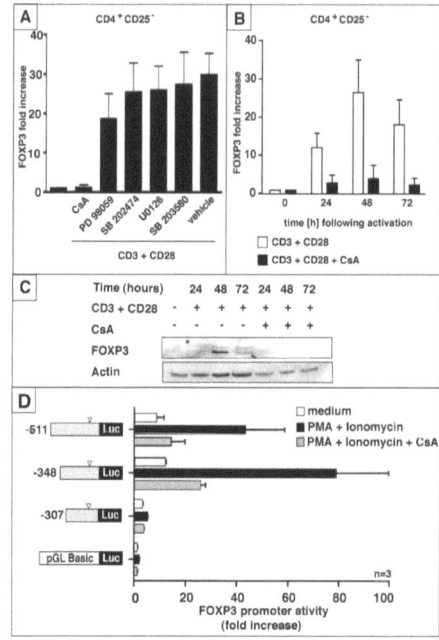

Figure 10: CsA inhibits FOXP3 induction in the CD4$^+$CD25$^-$ T cells. (A) CD4$^+$CD25$^-$ T cells were activated with anti-CD3 and anti-CD28 with CsA and different MAPK inhibitors. CsA potently inhibits FOXP3 induction. Bars show the mean ± SD of 3 independent experiments. (B) CD4$^+$CD25$^-$ T cells were activated with anti-CD3 and anti-CD28 with CsA, cells were harvested at different time points as indicated in the figure. Bars show the mean ± SD of 3 independent experiments. (C) Western blot analysis of FOXP3 in CD4+CD25- T cells after activation with anti-CD3 and anti-CD28 and with treatment with CsA (1 µM). Two independent experiments were done with similar results. (D) CD4$^+$CD25$^-$ T cells were transfected with the FOXP3 promoter constructs annd activated with PMA and ionomycin and treated or not with CsA.

NFAT and AP-1 are postive transactivators of FOXP3

Since the construct -348 was the shortest construct showing TCR responsiveness. We mutated the AP-1 binding sites, in the construct -348. Mutation of the AP-1 site at position -306, which is closest to the TSS has only a weak effect on promoter activity (Figure 11). In contrast, mutation of the AP-1 site located -324 strongly reduced the transactivational activity of the promoter (3-fold; Figure 11). The background was also reduced suggesting that those factors play an important role in the constitutive promoter activity.

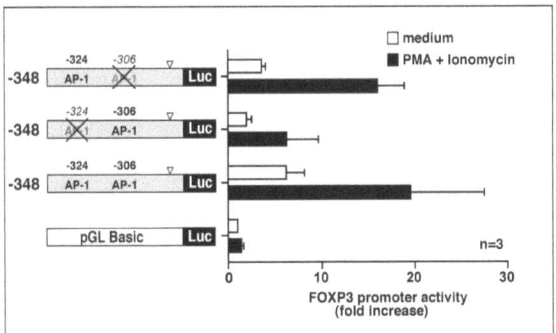

Figure 11: The AP-1 sites of the FOXP3 promoter have transactivatory activity. $CD4^+CD25^-$ T cells were transfected with construct containing AP-1 mutations and activated with PMA and ionomycin. AP-1 mutations decrease promoter activity, as well as its induction by activation. Bars show the mean ± SD of 3 independent experiments.

Mutations of the NFAT and AP-1 sites in the constructs -511 had a dramatic effect on the basal activity and inducibility by T cell activation (Figure 12A). Loss of the NFAT-490 and -328 in the construct -511 decreased the activity by 38 % and activation induces only 3 times instead of 4 times in the wild-type. Loss of the NFAT binding site –383 and AP-1 -476 decreases the activity by 55 % and the induction following activation was only of 2.2-fold (Figure 12A). The binding of NFATc2 to the NFAT sites on -490 and –328 was proven by pull-down assay, using cells lysates of activated $CD4^+$ T cells. NFATc2 was bound to the FOXP3 promoter oligonucleotides used for precipitation, but it was not precipitated by the mutated version (Figure 12B) or by the wildtype oligonucleotides competed by the excess of a NFAT consensus oligonucleotides (data not shown). To verify whether NFAT binds the FOXP3 promoter area on the chromatin under natural conditions, we performed

ChIP analysis. Starting from 2 hours after activation of CD4$^+$ T cells with anti-CD3 and anti-CD28 the binding of NFATc2 could be showed and was maximal after 5 hours and decreases thereafter (Figure 12D). Overexpression of NFATc2 dramatically increased promoter activity of the -511 construct (3-fold, relative to the empty pcDNA3 vector), which could be further increased by activation (6.5-fold; Figure 12E). The overexpression of NFATc2 had only a minor influence on the activity of the -90/+176 construct used as a control.

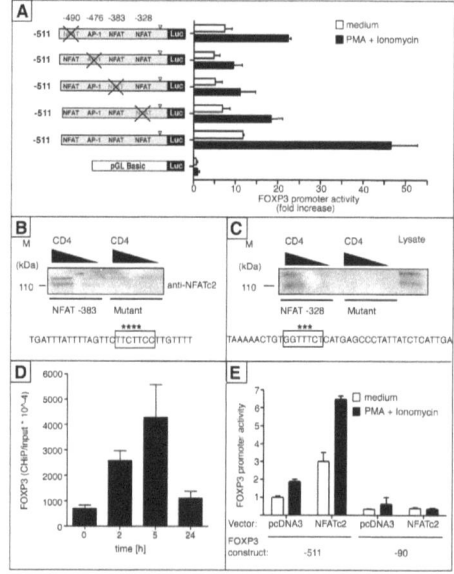

Figure 12: NFATc2 regulates human FOXP3 promoter activity in CD4$^+$CD25$^-$ T cells.
(A) Mutation of NFAT sites decrease promoter activity as well as its induction by activation by PMA and Ionomycin. Bars show the mean ± SD of 3 independent experiments. B) Nuclear extracts were prepared from CD4$^+$ T cells activated 2 hours with PMA and ionomycin. Biotinylated NFAT-376 and NFAT-328 oligonucleotides were absorbed by streptavidinagarose beads and then incubated with the nuclear extracts. Then the amounts of NFATC2 protein in the precipitates were assessed by immunoblotting with anti-NFATC2 mAb. Total nuclear extracts were also run as controls. Two independent experiments were done with similar results. (D) CD4$^+$CD25$^-$ T cells were activated using anti-CD3 and anti-CD28 and analyzed by ChIP for NFAT binding to the FOXP3 promoter. Quantitative fluorogenic PCR was performed. Data are expressed as the ratio of immuoprecipiated to input sequence and are mean of ± S.D. of two separate experiments. (E) Overexpression of CD4$^+$CD25$^-$ cells with NFATc2 with the 511 FOXP3 promoter construct increase the luciferase activity of the FOXP3 promoter constructs. NFATc2 could not further increase the –90 luciferase activity. Results shown are the mean ± S.D. of 1 experiment performed in triplicate. Two independent experiments were done with similar results.

Regulation of FOXP3 in CD4⁺CD25⁺ Tregs

It is known that overexpression of FOXP3 is sufficient to induce a T_{reg} phenotype, however the significance of FOXP3 regulation in already existing T cells is not clear. We therefore investigated whether activation has any effect on FOXP3 expression in Tregs. We activated FACS-sorted CD4⁺CD25⁺ Tregs or CD4⁺CD25⁻ effector T cells (Figure 13A) with plate-bound anti-CD3 and anti-CD28. After 3 days the cells were harvested and FOXP3 mRNA was measured by realtime-PCR. Resting CD4⁺CD25⁺ expressed about 90-fold more FOXP3 than CD4⁺CD25⁻ cells. Activation induced expression of FOXP3 mRNA by only 1.9-fold (Figure 13A), in contrast to 20-fold CD4⁺CD25⁻ cells (Figure 13B). In order to test whether this increase has a functional effect on Tregs function, we compared the suppressive capacity of unstimulated to preactivated Tregs. Although FOXP3 expression did only marginally increase (1.9-fold), the activation dramatically increased the suppressive capacity of Tregs (Figure 13C). However when Tregs were preactivated during 2 days in the presence of CsA, which was washed away before the cells were used in the suppression assay, the suppressive capacity was only marginally reduced (Figure 13C). Thus NFAT is important for FOXP3 induction, mediating regulatory differentiation, but does not affect the suppression of already existing Tregs, although activation potentates suppressive capacity.

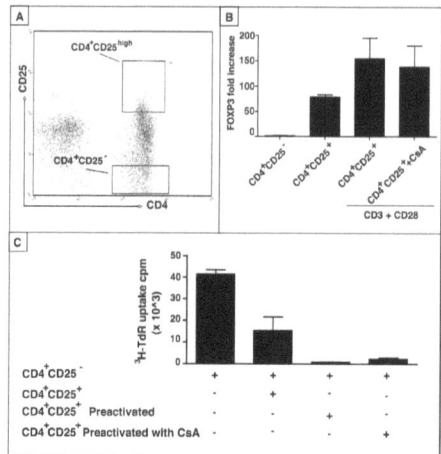

Figure 13: Activation does not induce FOXP3 expression in pre-existing Tregs. (A) CD4⁺CD25high were FACS-sorted using the shown gates. (B) FACS-sorted CD4⁺CD25⁻ Tregs were activated with anti-CD3 and anti-CD28 during 3 days and treated or not with CsA (1 µM). The cells were harvested for mRNA extraction. The percentage was calculated on the basis of the ΔΔCt method. Bars, 95% confidence interval calculated on the basis of deviation of EF-1 and FOXP3 expression. The results shown are the mean ± S.D. of three independent experiments. (C) Activation dramatically increases CD4⁺CD25⁺ Tregs suppressive capacity, CsA couldn't avoid this increase in the suppressive capacity. CD4⁺CD25⁺ Tregs were preactivated during 2 days in presence or absence of CsA. After washing the cells 3 times, their suppressive capacity on responder CD4⁺CD25⁻ was tested. 10 x 10⁴ CD4⁺CD25⁺ Tregs were added to 5 x 10⁴ CD4⁺CD25⁻. The results shown are the mean ± S.D. of three independent experiments.

Discussion

In the present study we describe the localization and structure of the human FOXP3 promoter, as well as elements, that are essential for its induction in T cells.

The FOXP3 promoter is located -6221 bp upstream of the translation start site and the 5'UTR is interrupted by a 6000 bp intron, which contains a splice donor site at the 5' end and a splice acceptor site at the 3' end, 22 bp upstream of the translation start site. The promoter is highly conserved between human, mouse and rat. The mRNA sequence published in the present study confirms the transcription start site and the location of the intron "0" of the reference sequence (NM_014009).

The chromatin accessibility is a key mechanism of gene regulation and has been shown to be essential for many genes during T cell differentiation like IL-4 and IFN-γ [109,274,275]. FOXP3 has been proposed to be a lineage-specific factor for Tregs and therefore the chromatin structure may be an important aspect of FOXP3 regulation [244,276]. FOXP3 was accessible in resting and activated $CD4^+CD25^-$ T cells, $CD4^+CD45RA$ and $CD4^+CD25^+$ but not in Jurkat and Hela cells, corresponding to their FOXP3 mRNA expression [254]. Thus chromatin remodeling may contribute to the cell-specific expression of FOXP3, controlling the access of the transcriptional machinery to the promoter. The $CD4^+CD25^-$ population showed an open chromatin conformation of FOXP3 gene, which was further increased by activation. The non-repressive chromatin configuration may therefore allow $CD4^+CD25^-$ T cells to acquire a regulatory phenotype upon activation with the appropriate key of transcription factors.

In order to identify this set of transcription factors we analyzed the 1.6-kbp region upstream of the TSS. This region showed promoter activity, when cloned in front of a luciferase reporter gene and transfected into primary $CD4^+$ T cells. In contrast, Hela and CHO cells did not show any promoter activity. Thus FOXP3 cell-specificity is regulated not only at the chromatin, but also on transcription level. The serial deletion constructs revealed that a fragment of 348 bp contained the minimal promoter necessary for the induction of the gene. The deletion of 245 bp upstream of the TSS totally abrogated the promoter activity, indicating that this area contains the core promoter. The current data show that the specific mutation of the TATA (-34), the GC (-138) and CAAT boxes (-218), reduce activity of the core promoter. Furthermore, we demonstrate that the GC box is in fact bound by Sp1 and Sp3. Since these factors are characteristic for eukaryotic promoters [277], these data confirm the location of the FOXP3 promoter.

On the basis of these results we analyzed inducible elements upstream of this area. We demonstrate that FOXP3 expression is induced following TCR engagement in $CD4^+CD25^-$ T cells. Activation of $CD4^+CD25^-$ T cells with anti-CD3 or PMA and ionomycin induced FOXP3 promoter activity in the -511 reporter gene. This result shows that TCR-engagement acts directly on the FOXP3 promoter and confirms previous studies [63] showing that in vitro activation of $CD4^+CD25^-$ cells was sufficient to generate cells expressing FOXP3, which have suppressive capacity. In fact, exposure to an antigen [278], TGF-β [210,230,265], estrogen [279,280] or glucocorticoids [262] along with T cell activation can induces FOXP3 in $CD4^+CD25^-$ T cells. Therefore, activation seems to be a key event in the generation of Tregs, as it was previously shown to be essential in the differentiation process of Th1 and Th2 cells [281-283].

We narrowed down the activation dependence to the minimal FOXP3 promoter (-348), whereas the fragment that is just 41 bp shorter does not show any induction. Therefore the activation-responsive element of the FOXP3 promoter is located between -511 and -307. NFAT and AP-1 are well known mediators of T cell activation and are clustered in this region. Mutations disrupting the NFAT and AP-1 binding sites decreased the luciferase activity, revealing their role in the transactivation of the FOXP3 promoter. The activation of the FOXP3 gene is mediated by at least three NFAT sites, which we demonstrated to be bound by NFATc2 and three AP-1 sites, in proximity of NFAT sites. Those transcription factors often cooperate to induce cytokine gene expression and are forming complexes as in the promoter of IL-2 [284-286], IL-4 [287], IFN-γ [288] and CTLA-4 [289].

The MAPK-inhibitor (PD98059) only partially inhibited activation-induced FOXP3 mRNA expression, suggesting that the AP-1 factors can be mobilized by others pathways. In contrast, CsA completely inhibited the mRNA induction of FOXP3 as well as the promoter activity. CsA is a well-known immunosuppressive drug which, blocks NFAT translocation into the nucleus by inhibition calcineurin phosphatase activity [290]. We have previously shown that immunosuppressant glucocorticoids promote FOXP3 expression [262], whereas rapamycin does neither enhances nor decreases FOXP3 (data not shown and [291,292]). Therefore immunosuppressive drugs may have different mechanisms to promote tolerance induction.

Alternatively, immunosuppressive drugs may also act on pre-existing Tregs that are only marginally affected by TCR engagement in terms of FOXP3 mRNA expression, which is already high in resting Tregs. This marginal enhancement of FOXP3 expression in already existing, anergic Tregs is resistant to CsA expression, confirming previous studies, showing that anergic cells are impaired in Ca^{++}/NFAT mobilization [293,294]. However, pre-activation shows a dramatic increase on the suppressive capacity of the $CD4^+CD25^+$ Treg cells. Pretreatment of the $CD4^+CD25^+$ cells with CsA had just a minor effect on the suppressive capacity and suggest that NFAT is not essential in the process of suppression. In fact, $CD4^+CD25^+$ T_{regs} have been shown to be anergic and hyporesponsive to TCR stimulation and unable to induce Ca^{++} signaling, that may explain that those cells are unable to further induce FOXP3 expression upon activation [295].

Taken together our results indicate that the FOXP3 promoter is cell-specific and is active only in primary T cells. The identified basal promoter has similarities to immunological genes carrying elements including NFAT and AP-1, which are induced following TCR engagement. The reporter-constructs provide new tools to identify mechanisms underlying tolerance induction and potential therapeutic interventions.

Acknowledgements

We thank Prof G. Suske (University of Marburg, Germany) for providing us Sp1 and Sp3 antiserum. We thank Prof. A. Rao (Harvard Medical School, Boston, USA) for providing the NFATc2 construct.

Table I. *Primer.* Primers FoxP3 rev, 1657, -1210, -511, -307 and -211 introduced a restriction enzyme recognition site for KpnI or XhoI. The primers used for mutational analysis are also shown. The underlined letters denote mutated nucleotides.

Name	Sequence (5' → 3')	Purpose a
FOXP3 +176 rwd	AACTCGAGACCTTACCTGGCTGGAATCACG	Cloning
Fox-1657	AAGGTACCCTTGGCCACCAGATTTGTACC	Cloning
Fox-1210	AAGGTACCCTACCTCCGTTTCCCTCATCTG	Cloning

Fox −511	AAGGTACCTTCCCATCCACACATAGAGC	Cloning
Fox-307	AAGGTACCATACCTCTCACCTCTGTGGTG	Cloning
Fox −211	AAGGTACCAGTCTCATAATCAAGAAAAGG	Cloning
TATA −34	GCGTGGTTTTTCTTCTCGGTCTCGAAGCAAAGTTGTTTTTGATACG	Mutatic
GC Sp1 - 142	GAGAGAAAAAAAAAACTATGAGAACCTTTTCCCACCCCGTGATTATCAGCGC	Mutatic
NFAT -328	CTATACACTTTTGTTTTAAAAACTGTGGGAGCTCATGAGCCCTATTATCTCATTGATACC	Mutatic
NFAT -490	CATAGAGCTTCAGATTCTCTTTCTTGGACCAGAGACCCTCAAATATCCTCTCAC	Mutatic
AP-1 -476	TCATGAGCCCTATTATCTCCACGATACCTCTCACCTCTGTGG	Mutatic
NFAT −383	GTTGGCCCTGTGATTTATTTTAGTTCTCGAGCCTTGTTTTTTTTTTTCAAACTCTATACAC	Mutatic
anchor primer	GGCCACGCGTCGACTAGTACGGGIIGGGIIGGGIIG	RACE
RACEFOXP3+987	CACCCGCACAAAGCACTTG	RACE
RACEFOXP3+521	GCTGCTCCAGAGACTGTACCATCT	RACE

2.2. GATA3 driven Th2 responses inhibit FOXP3 expression and the formation of regulatory T cells

Pierre-Yves Mantel[1], Harmjan Kuipers[2], Onur Boyman[3], Nadia Ouaked[1], Beate Rückert[1], Christian Karagiannidis[1], Bart N. Lambrecht[2], Rudolf W. Hendriks[4], Kurt Blaser[1], Carsten B. Schmidt-Weber[1,5]

1) Swiss Institute of Allergy and Asthma Research Davos (SIAF), CH-7270 Davos-Platz, Switzerland
2) Department of Pulmonary Medicine, Erasmus MC, Dr Molewaterplein 50, 3015 GE Rotterdam, The Netherlands
3) Division of Immunology and Allergy, University Hospital of Lausanne (CHUV), CH-1011 Lausanne, and Ludwig Institute for Cancer Research, University of Lausanne, CH-1066 Epalinges, Switzerland
4) Department of Immunology, Erasmus MC, Dr Molewaterplein 50, 3015 GE Rotterdam, The Netherlands
5) corresponding author: csweber@siaf.unizh.ch

Submitted to Science

Summary

The origin of regulatory T cells (Tregs), controlling inflammatory responses against autoantigens or allergens is unknown. The present study describes a mechanism repressing peripheral Treg induction on the basis of the FOXP3 promoter analysis and challenges the idea of a shared developmental pathway with T helper type I (Th1) or Th2 lineages. Instead present data favor the concept of a distinct lineage whereby Tregs develop alternatively to Th1 or Th2 cells. We demonstrate that cytokines such as IL-4 and TGF-β present at the time of T cell priming of the uncommitted cells are decisive not only in differentiating T cells towards effector phenotypes, but also towards Tregs. Moreover, Th2-driving conditions that occur in allergic inflammation prevent the induction of Tregs by a GATA3-mediated inhibition of the FOXP3 promoter. Since IL-4 treatment in mice reduces Treg frequency, therapeutic approaches targeting IL4 and/or GATA3 might provide new preventive strategies facilitating tolerance induction.

Introduction

Effective immune responses are characterized by T cell activation, which directs adaptive and innate immune responses to efficiently kill pathogens. Dependent on the pathogen, T cells differentiate into different subtypes such as Th1 or Th2 cells, which are most efficient in defeating microbial or parasitic invaders respectively. The balance between Th1 and Th2 cells has been the starting point for therapeutic interventions [296]. A hallmark of Th1 and Th2 differentiation pathways is the exclusiveness of the individual mechanisms. IL-12 mediated STAT4-phosphorylation [102] and T-bet expression are essential for Th1 differentiation [297,298]. In contrast, IL-4-induced STAT6 and GATA3 inhibit differentiation into Th1 cells in the early phase of commitment [85,101]. GATA3 is sufficient to induce Th2 phenotypes [299] and acts not only through induction of IL-4, IL-5 and IL-13, the Th2 cytokines, but also through inhibition of Th1 cell-specific factors [299]. Recently it could be shown that T-bet directly modulates GATA3 function [300], suggesting that transcription factors compete in the early differentiation phase of T cells to finally imprint the T cell phenotype [301]. A GATA3 dominated immune response has been shown to be essential for airway hyperresonsiveness [302] and can break antigen-specific immune tolerance [303]. Overexpression of a dominant negative form of GATA3 [304] or treatment with antisense-mediated GATA3 blockade [305] decreased the severity of the allergic airway hyperresponsiveness.

The discovery of Tregs highlights another phenotype of T cells, which is essential for tolerance against autoantigens. However its integration in lineage development is not fully clear. Naturally-occurring Tregs (nTregs) are generated in the thymus and are assumed to protect against the activity of autoreactive T cells in the periphery. These cells express the forkhead transcription factor FOXP3 and constitutively express CD25 on their surface, but lack cytokine expression, which would set them in proximity of Th1 or Th2 lineages. Particularly interesting are T_{regs}, which are generated in the periphery and thus are potential targets for therapeutic intervention. These induced Tregs (iTregs) were reported to express FOXP3, however expression may be transient [306]. The exact circumstances of iT_{reg} generation are unclear, but TGF-β has been demonstrated to be important for the induction of these cells *in vitro* and *in vivo*, since animals lacking the TGF-βRII on T cells are deficient in peripherally-iT$_{regs}$ and suffer from a T cell dependent multiorgan inflammatory disease [64,191,210,307,233]. Although the effect of TGF-β on Treg induction is well documented, its molecular targets remain to be identified.

Two different scenarios of Treg induction can be hypothesized: one, in which Tregs can be induced in already committed effector cells, which provides a scenario of an inherent shutdown mechanism of T cell activation. The other suggests that Tregs differentiate from naïve T cells as a separate lineage in a similar fashion as known for Th1 or Th2 commitment, a scenario providing a suppressive memory population. The current study provides evidence for the second model and focusses on GATA3 and FOXP3 which may play a similar role in Treg commitment as T-bet and GATA3 for Th1 and Th2 differentiation respectively. This assumption is based on the observation that TCR activation is necessary to generate Tregs as well as the FOXP3 gene and that high and stable FOXP3 expression is sufficient to generate a regulatory phenotype [240-242]. We show that GATA3 excludes FOXP3 expression and that IL-4, and thus Th2 cells inhibits FOXP3 expression both *in vitro* and *in vivo*. We demonstrated that GATA3 directly binds to the FOXP3 promoter and thereby inhibits the induction of this gene, supporting a concept of a separate cell lineage for T$_{reg}$ commitment.

Results

Exclusive commitment of Tregs

To investigate whether FOXP3 can be expressed by any T cell subset or is restricted to a distinct lineage, FOXP3 mRNA expression was analyzed in freshly isolated T cell such as CD25-depleted $CD4^+$cells, $CD45RA^+$ naïve or $CD45RO^+$ memory T cells as well as T cells driven *in vitro* towards Th1, Th2 or iT_{reg} cells (Figure 14A). The $CD4^+CD25^-$, $CD45RA^+$, $CD45RO^+$ and $CD4^+CD45RO^+CD25^-$ were able to significantly induce FOXP3 mRNA up to 40 - 50-fold upon TCR activation and addition of TGF-β. Th1 cells showed only a 10-fold increase. In contrast, Th2 cells stimulated under the same conditions did not increase FOXP3 expression. The *in vitro* iTregs were unable to further upregulate FOXP3, which was already high under the resting conditions (Figure 14A).

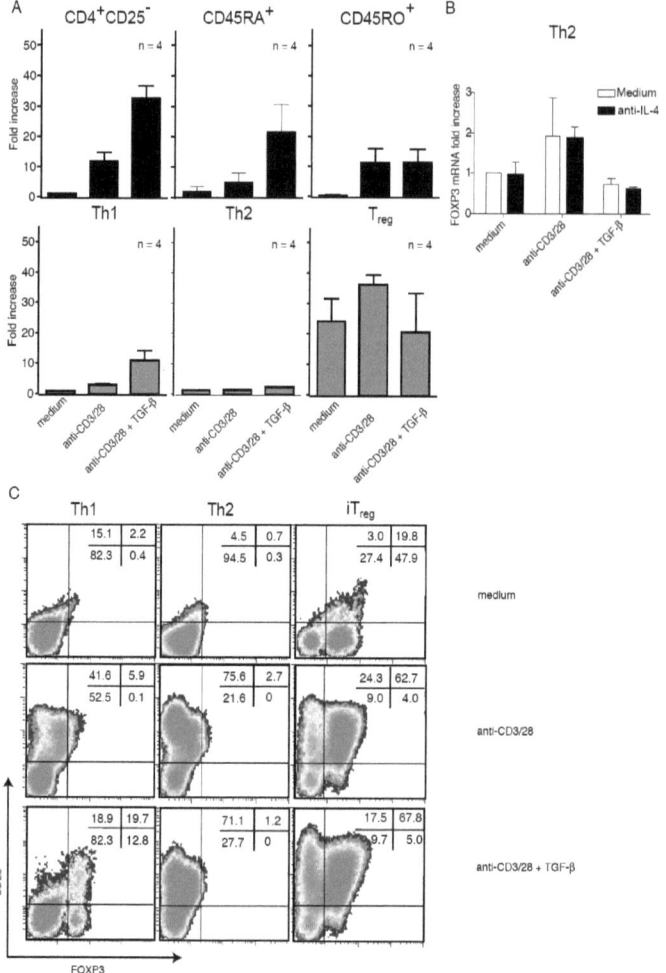

Figure 14. Th2 cells cannot induce FOXP3 expression
(A) Human T cells were activated with plate-bound anti-CD3/CD28 with or without TGF-β. Cells were harvested after 5 days and FOXP3 mRNA was quantified by real-time PCR. Bars show the mean ± SD of 4 independent experiments.
(B) Th2 cells were activated with anti-CD3/CD28, TGF-β or anti-IL-4 as indicated. Bars show the mean ± SD of 3 independent experiments.
(C) FACS analysis of intracellular FOXP3 expression of Th1, Th2 or i Treg differentiated (2 rounds) cells in activated or resting conditions in the presence of TGF-b. FOXP3 expression was measured after five days in culture. The dot blots are representative of three independent experiments.

Th2 cells are known to produce IL-4 upon activation, which may interact with TGF-β signaling and thus prevent FOXP3 induction. However, the neutralization of IL-4 with a blocking IL-4 antibody, did not rescue FOXP3 expression in the differentiated Th2 cells. These data demonstrated that Th2 have a limited capacity to express FOXP3 (Figure 14B).

The inability of Th2 cells to express FOXP3 was also documented at the single cell level, revealing that Th2 cells lack FOXP3 expression under any condition, confirming the mRNA analysis (Figure 14C). Only iTregs expressed FOXP3 in resting conditions. Interestingly, we observed that resting iTregs express FOXP3, but show low CD25 surface expression. Repeated exposure to TGF-β did not further increase the FOXP3 expression in the iTreg lineage, but induced transiently FOXP3 expression in Th1 cells.

Taken together these results indicate that FOXP3 can be induced in naïve T cells, but not in commited Th2 cells, which may lack factors necessary for FOXP3 expression or have an inhibitory mechanism that affect their ability to express FOXP3.

FOXP3 and GATA3 kinetic in differentiating cells

The limited capacity of differentiated effector cells to induce FOXP3 expression, suggested that Treg induction has to occur before differentiation. Following initiation of the differentiation process FOXP3 and GATA3 show a similar activation kinetic within the first three days, which are considered to be critical in commitment [275]. Under Th2 differentiation conditions FOXP3 expression increased marginally during the first hours and then decreased to background levels (Figure 15A). The Th0 and Th1 cells expressed a higher level than the Th2, but lower than the iTregs (data not shown). Thus, although GATA3 and FOXP3 show similar kinetics, their expression is mutually exclusive.

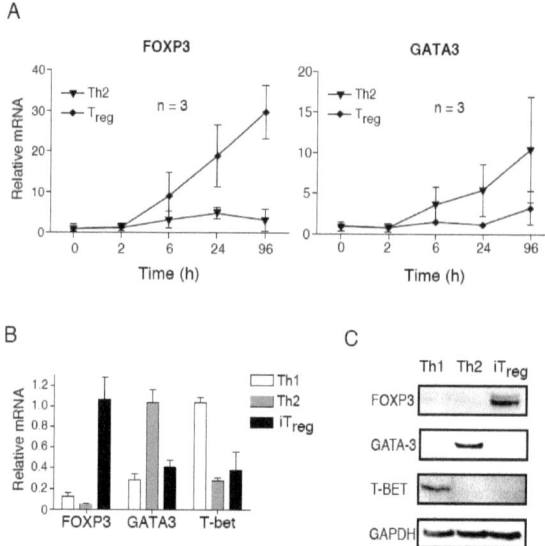

Figure 15: FOXP3 induction during the differentiation process.
(A) CD4⁺CD45RA were activated with plate-bound anti-CD3/CD28 in the presence of TGF-β (5 ng/ml)or IL-4 (25 ng/ml). The cells were harvested at different time-points and mRNA was quantified by real-time PCR for FOXP3 and GATA3 expression. Bars show the mean ± SD of 3 independent experiments.
(B) and (C) *in vitro* differentiated Th1, Th2 or iTregs were restimulated with plate-bound anti-CD3/CD28. Cells were harvested after 3 days for real-time PCR (B) or western blot analysis (C) of FOXP3, GATA3 and T-bet; GAPDH served as internal control. Data are representative of three independent experiments.

The selective expression of FOXP3 was even more apparent three days after restimulation of the cell lineages, where FOXP3 mRNA and protein were exclusively detected in Tregs (Figure 15B and C). These cells displayed phenotype comparable to nTregs, including an anergic phenotype upon anti-CD3 re-stimulation, CD103, CTLA-4, GITR and PD-1 surface expression (Figure 16A and B). The iTregs neither produced the Th1 cytokines IFN-γ nor the Th2 cytokines IL-4 or IL-13, upon activation with PMA and ionomycin, as determined by FACS (Figure 17) and ELISA (data not shown).

A

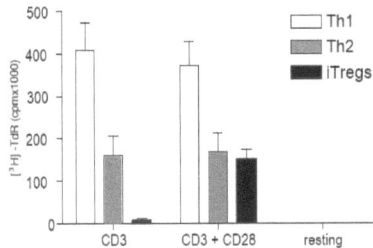

B

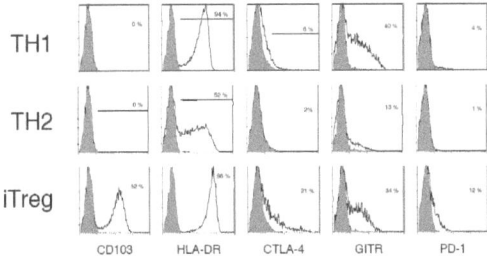

Figure 16: iT_regs have a different phenotype than the Th1 and Th2.
(A) *in vitro* differentiated Th1, Th2 or iTregs were stimulated as indicated and proliferation was measured after 3 days by thymidine incorporation overnight. Bars show the mean ± SD of 3 independent experiments.
(B) Surface markers were measured by flow cytometry. The data are representative of three independent experiments.

These data demonstrated that FOXP3 and thus Tregs are in fact co-evolving with Th1 and Th2 cells as a separate lineage.

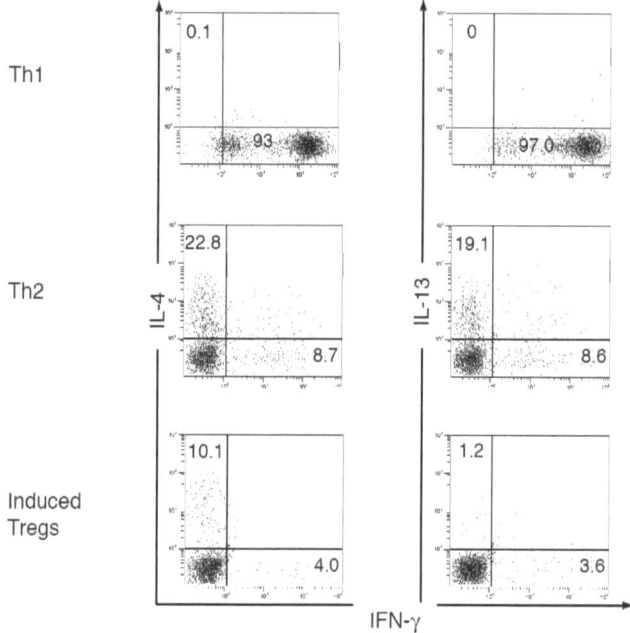

Figure 17: IL-4, IFN-γ and IL-13 cytokine production by Th1, Th2 and iTregs.
Differentiated T cells were stimulated with PMA and ionomycin during 4 hours and subjected to intracellular staining by FACS. The data are representative of three independent experiments.

IL-4 inhibits TGF-β-mediated T_{reg} commitment

To characterize the molecular nature of T_{reg} commitment, we investigated GATA3 and FOXP3 expression in the presence of IL-4 and TGF-β in the early differentiation phase of T cells into Th2 and T_{reg} conditions. Human $CD4^+CD45RA^+$ T cells were activated with plate-bound anti-CD3/CD28 in the presence of TGF-β and/or IL-4 and harvested after 5 days. IL-4 efficiently repressed the TGF-β-mediated induction of FOXP3 expression (Figure 18A) in a dose-dependent manner (Figure 18B). Of note, GATA3 was also induced in the presence of TGF-β at high IL-4-concentration (Figure 18A and B). These IL-4-treated cells (Figure 18A) were not able to suppress proliferation of autologous target T cells in an *in vitro* suppression assay (data not shown). The IL-4-mediated prevention of FOXP3 expression was not caused by interferences of the receptor signalling, since the phosphorylation of SMAD2 or

STAT6 was not affected by the addition of IL-4 and/or TGF-β, demonstrating that IL-4 as well as TGF-β signalling were functional under these conditions (Figure 18C). Furthermore IL-4 was inhibiting TGF-β-mediated generation of FOXP3$^+$ T cells, as shown by FACS analysis (Figure 18D).

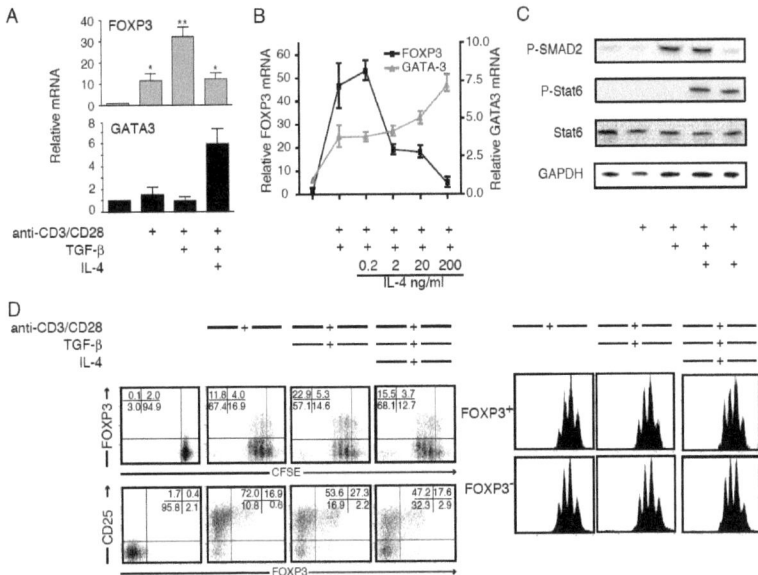

Figure 18: IL-4 inhibits TGF-β mediated Treg commitment.
(A) CD4$^+$CD45RA$^+$ were activated with plate-bound anti-CD3/CD28 during five days in presence of TGF-β (10 ng/ml) and with or without IL-4 (100 ng/ml). The cells were harvested and mRNA was quantified by real-time PCR for FOXP3 and GATA3 expression. Bars show the mean ± SD of 6 independent experiments. Statistical analysis was performed using the Dunnett test. Statistical significance is indicated by asterisks (* = p≤ 0.05, ** = p≤ 0.01).
(B) CD4$^+$CD45RA$^+$ cells were activated in the presence of a constant concentration of TGF-β (5 ng/ml) with an increasing concentration of IL-4, as indicated. Cells were harvested after five days for mRNA quantitation.
(C) CD4$^+$CD45RA$^+$ cells were stimulated in vitro with plate-bound anti-CD3/CD28, TGF-β (10ng/ml) and IL-4 (100 ng/ml) as indicated. After 1 h, cell lysates were prepared and analyzed by western blot for phosphorylated SMAD2 and STAT6. Total STAT6 and GAPDH served as internal control.
(D) CFSE-labeled CD4$^+$CD45RA$^+$ were activated with plate-bound anti-CD3/CD28, TGF-β and IL-4, as indicated. After 5 days, cells were analyzed by flow cytometry. CFSE profiles, surface CD25 and intracellular FOXP3 expression are shown on the left panels. On the right comparison of CFSE profiles for FOXP3$^+$ and FOXP3$^-$ cells is shown for the conditions as described. Data are reprevative of four independent experiments.

It is known that IL-4 is a potent growth factor and may therefore favor the proliferation of (FOXP3⁻) effector cells and thus decrease the relative percentage of FOXP3⁺ cells. However, analysis of cell division kinetics by CSFE labelling demonstrated that IL-4 did not differentially promote cell growth of FOXP3⁺ and FOXP3⁻. In fact both populations showed similarly enhanced proliferation and more cells were able to reach three cycles of division (Figure 18D, right panel). Furthermore the TGF-β-mediated induction of FOXP3 expression was not caused by overgrowth of a CD25⁻FOXP3⁺ minority, since the number of FOXP3⁺ cells was low/absent in the purified CD4⁺CD45RA⁺ T cells (between 0 and 1 %) and the FOXP3⁺ cells were not confined to the highly divided cells. The FOXP3⁺ T cells were generated from FOXP3⁻ T cells, since level of non-dividing FOXP3⁺ cells increased, revealing FOXP3⁺ induction out of FOXP3⁻ cells (Figure 18D, left panel). CD25 was downregulated in TGF-β-treated cells compared to activated T cells, which was even more pronounced in cells cultured with TGF-β and IL-4. The FOXP3⁺ T cells are shown to be CD25⁺ at this early stage of differentiation (Figure 18D left panel).

These results show that IL-4 acts *in vitro* as an inhibitor of FOXP3 expression, without interfering with TGF-β signaling, probably acting at the level of transcription factors.

GATA3 is a negative regulator of FOXP3 expression
The data presented above showed that IL-4 potently represses T_{reg} commitment without affecting TGF-β signaling. GATA3 was previously shown to repress Th1 commitment by inhibiting STAT4 expression [101,102] and therefore to prevent differentiation into Th1 cells. Thus we hypothetized a potential role for GATA3 in repressing FOXP3 and therefore preventing the commitment to T_{regs}. To see whether GATA3 can directly inhibit FOXP3 induction, we overexpressed GATA3 in human primary CD4⁺CD45RA⁺ T cell. After transfection, the cells were activated with soluble anti-CD3/CD28 in the presence or absence of TGF-β. GATA3-transfected cells showed a lower FOXP3 mRNA level when treated with TGF-β compared to the cells transfected with the control vector (Figure 19A, left panel). Successful GATA3 expression was controlled by western blot analysis (Figure 19A, right panel). This result revealed that GATA3 directly inhibits the Foxp3 gene. Since transient overexpression may be limited by time and transfection efficiency, we analyzed the

inhibitory effect of GATA3 on FOXP3 in transgenic DO11.10 mice, overexpressing GATA3 under the control of the human CD2 locus control region (DO11.10xCD2-GATA3). This model provides OVA-specific CD4$^+$ T cells expresssing GATA3 constitutively. The frequency of peripheral natural occuring T regs (CD4$^+$CD25$^+$Foxp3$^+$) in these mice was slightly decreased compared to DO11.10 littermate control mice (Figure 20). Furthermore thymic selection into the CD4 lineage is largely intact in DO11.10 x CD2-GATA3 (R.W.Hendriks, unpublished data). These mice develop thymic lymphomas at older age, but signs of autoimmune disease were not described [308]. To investigate the effect of GATA3 on iT$_{regs}$, CD4$^+$CD62L$^+$CD25$^-$ cells were isolated, activated with OVA in the presence or absence of TGF-β and Foxp3 expression was analyzed after four days. The naïve CD4$^+$CD25$^-$ cells were Foxp3$^-$ (data not shown) and activation with the OVA antigen did only marginally induce Foxp3$^+$ cells. As described for the human cells, TGF-β dramatically upregulated Foxp3 in the DO11.10 littermate control mice. In contrast, cells from the CD2-GATA3xDO11.10 mice were almost unable to upregulate Foxp3, when activated with TGF-β and OVA (Figure 4B). All mice produced similar amounts of TGF-β and Smad7 was equally expressed [309] in T cells of both mice strains (Figure 19C), indicating intact TGF-β signaling. Taken together these results demonstrated a repressive role of IL-4-induced GATA3 transcription factor in the generation of iTregs.

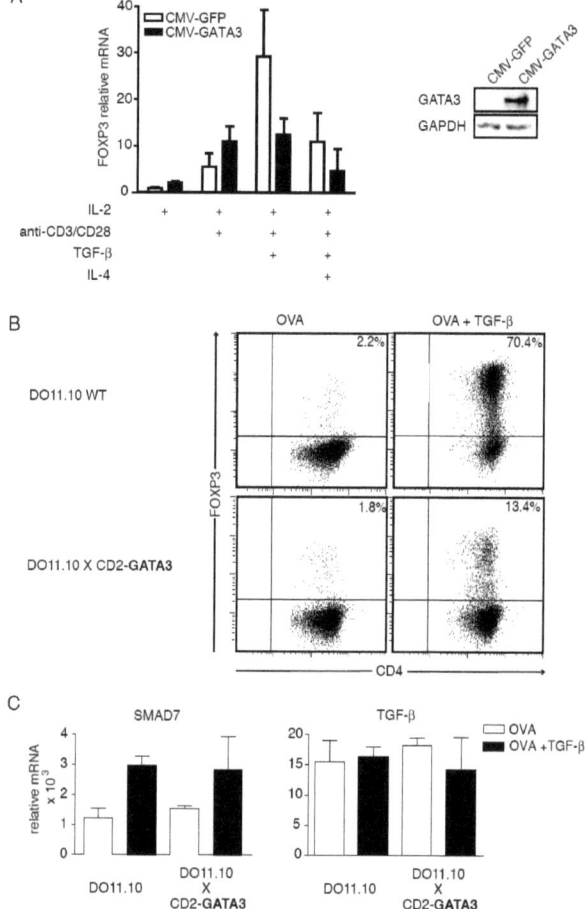

Figure 19: GATA3 acts as a negative regulator of FOXP3 expression.
(A) human CD4+CD25- cells were transfected with a GATA3 or GFP (as negative control) expression construct and treated with anti-CD3/CD28, and TGF-β, as indicated. The mRNA was analyzed after 3 days by real-time PCR.
(B) CD4+CD25- cells were isolated from DO11.10 and DO11.10xCD2-GATA3 mice and treated with OVA and TGF-β for 96 h. Surface CD4 and intracellular Foxp3 were measured by FACS. These data are representative of 3 different experiments.
(C) The cells treated as in (B) were harvested and mRNA was quantified by real-time PCR for SMAD7 and TGF-β expression. Bars show the mean ± SD of 3 independent experiments.

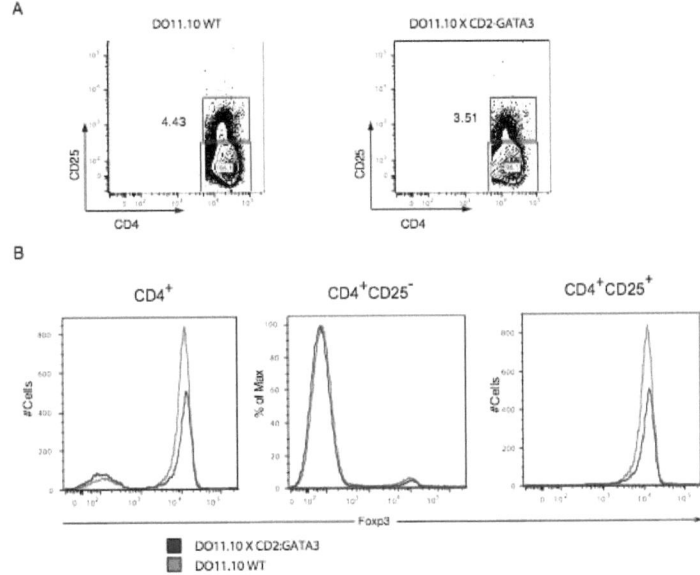

Figure 20: CD2-GATA3xDO11.10 CD4⁺Tcells do have Foxp3⁺ T cells.
(A) $CD4^+$ T cells from CD2-GATA3xDO11.10 or DO11.10 WT mice were isolated from pooled lymph nodes and spleens and analyzed by FACS for $CD25^+$ and $CD4^+$ cells
(B) $CD4^+CD25^-$ and $CD4^+CD25^+$ from CD2-GATA3xDO11.10 or DO11.10 WT were analyzed for Foxp3 expression. The CD2-GATA3xDO11.10 show a reduced number of $FOXP3^+$ cells in the $CD4^+CD25^+$ population. The data are representative of three independent experiments.

GATA3 directly binds to and represses the FOXP3 promoter

To investigate the molecular mechanism of GATA3 mediated repression of FOXP3, the human FOXP3 promoter was studied and a palindromic binding site for GATA3 was discovered. The GATA-binding site is located -400 bp upstream from the transcription start site, in a region which has already been described as important for the regulation of FOXP3 expression [310]. This site is highly conserved between human, mice and rat (Figure 21) and may therefore play an important role in FOXP3 regulation. The functional relevance of this site was studied using an established FOXP3-promoter construct [310]. We transfected human primary $CD4^+$ T cells and Jurkat cells, the latter constitutively expressing GATA3 [311,312] and measured FOXP3 promoter activity. The promoter was not active in the GATA3-expressing cell line Jurkat, whereas the construct was active in the CD4 cells, which express a lower amount of GATA3 (Figure 22A). A site-specific mutation abolishing the GATA3-

binding site of the human FOXP3 promoter increased luciferase activity by 2.5-fold in CD4⁺ T cells, revealing a repressor activity of GATA3 on the FOXP3 promoter (Figure 22B). Overexpression of GATA3 diminished luciferase activity of the FOXP3 promoter compared to the control vector (pcDNA3; Figure 22C).

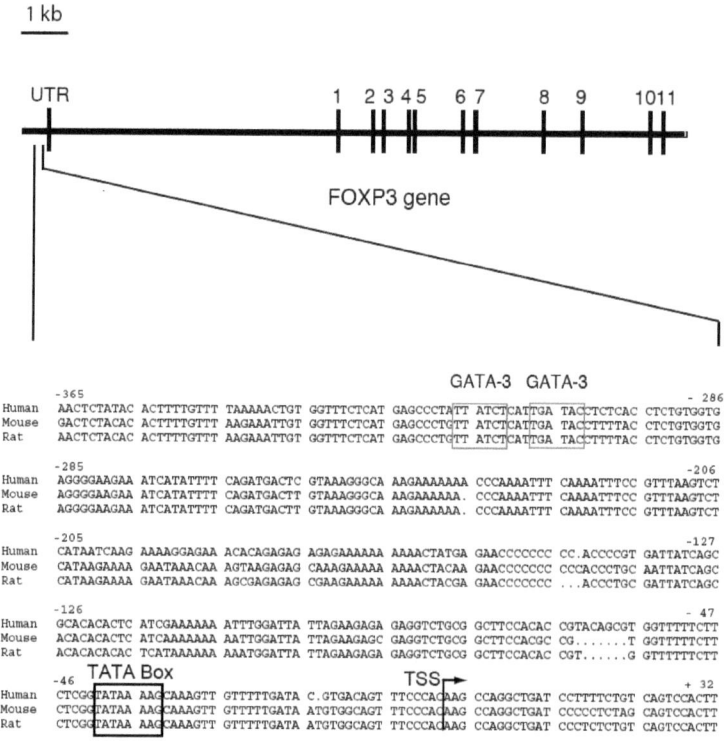

Figure 21: Localisation of the human FOXP3 promoter in a highly conserved region of the FOXP3 gene.
Structure of the FOXP3 gene and sequence conservation of the human (top: GenBank accession number AF235097), mouse (middle: accession number AF277994) and rat (bottom: GenBank accession number NW_048035) are shown. The transcription start site (TSS) is indicated by an arrow. GATA3-binding sites are indicated by a box.

The ability of GATA3 to physically interact with the FOXP3 promoter was further investigated. Cells which normally do not express GATA3 (HEK) were transiently transfected with GATA3- or control-genes and increasing amounts of lysates were incubated with oligonucleotides containing the GATA3-site of the FOXP3 promoter or a control oligonucleotide with a mutated GATA3 binding site. These oligonucleotides were precipitated and GATA3-specifically detected by western blot. Similarily, GATA3-expressing Th2 cells and iTregs were subjected to this approach. Only HEK cells overexpressing GATA3 (Figure 22D) and Th2 cells (Figure 22E) showed GATA3-binding activity. This experiment proves that GATA3 can bind the FOXP3 promoter, but leaves open, whether this binding activity also occurs in an intact T cell. Therefore we performed a chromatin coimmunoprecipitation (ChIP) using a GATA3-specific antibody to precipitate chromatin of Th2 cells and iTregs. In line with the previous experiments, we could detect GATA3 binding to the FOXP3 promoter in Th2 cells, but not in the iTregs (Figure 22F).

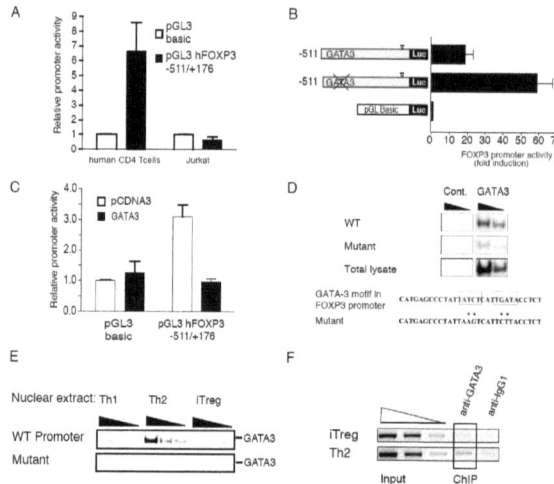

Figure 22: GATA3 binds to and represses the human FOXP3 promoter.
(A) Jurkat and human primary CD4 cells were transfected with an empty vector (pGL3 basic) or vector containing the putative FOXP3 promoter region. Bars show the mean ± SD of arbitrary light units normalized for renilla luciferase of 4 independent experiments; samples were measured as triplicates.
(B) CD4 cells were transfected with wild-type or a GATA3 mutated FOXP3 promoter reporter construct. Bars show the mean ± SD of 3 independent experiments.
(C) Overexpression of GATA3 in CD4 cells with the 511 FOXP3 promoter construct decreases the luciferase activity of the FOXP3 promoter. Results shown are the mean ± S.D. of 3 different experiments performed in triplicate.
(D) Nuclear extracts were prepared from HEK cells transfected with GATA3 or an empty vector. Biotinylated oligonucleotides were absorbed by streptavidin agarose beads and then incubated with nuclear extracts. The amounts of GATA3 protein in the precipitates were assessed by immunoblotting with anti-GATA3 mAb. Total nuclear extracts were also run as controls. These data are representative of 3 different experiments.
(E) Nuclear extracts from Th1, Th2 and iTregs were collected and tested for GATA3 binding activity as in (D). This experiment is representative of 3 experiments. (F) Th2 and iTregs were analyzed by ChIP for GATA3 binding to the FOXP3 promoter. Shown is the PCR for the FOXP3 gene after reversing the cross-linking. The "input" represents PCR amplification of the total sample, which was not subjected to any precipitation. Results are representative of three independent experiments.

Taken together these data demonstrated that GATA3 directly binds the FOXP3 promoter and inhibits its activity.

IL-4 does not affect FOXP3 expression of existing, terminally differentiated T_{regs}.

The dramatic effects of GATA3 in preventing Treg commitment may also alter the FOXP3 expression and suppressive function of already existing, terminally differentiated Tregs. Accordingly terminally-differentiated iTregs as well as natural $CD4^+CD25^{high}$ Tregs were activated and treated with IL-4. In contrast to the potent effect of IL-4 on differentiating cells, IL-4 was not able to decrease FOXP3 mRNA expression in either the natural (Figure 23A) or on existing iTregs (Figure 23B). Furthermore pre-activation of natural Tregs with anti-CD3/CD28 and IL-4 did not diminish the suppressive capacity of these cells (Figure 23C). These results reveal that GATA3 inhibits Tregs primarily during the differentiation process.

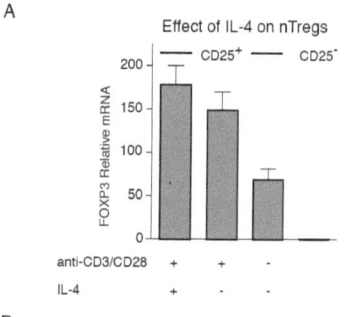

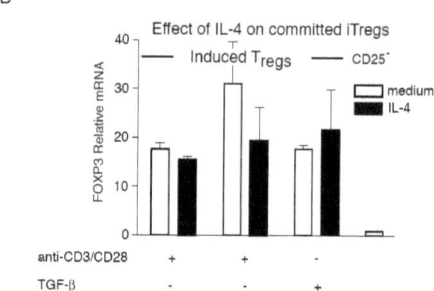

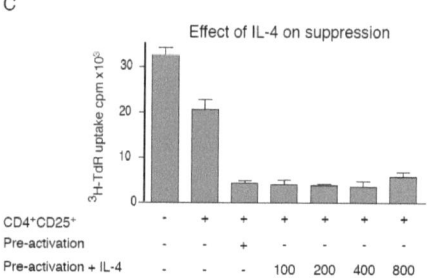

Figure 23: IL-4 does not revert T_{reg} phenotype in vitro.
(A-G) $CD4^+CD25^{high}$ cells were FACS-sorted and activated with plate-bound anti-CD3/CD28 plus IL-2 during 3 days and in the presence or absence of IL-4 (100 ng/ml) and harvested for real-time PCR analysis. The results shown represent the mean ± S.D. of three independent experiments.
(B) T_{regs} were induced in vitro and treated as described in (A).
(C) Activation dramatically increases $CD4^+CD25^+$ Tregs suppressive capacity of $CD4^+CD25^+$ Tregs. $CD4^+CD25^+$ T_{regs} were preactivated during 2 days in the presence or absence of an increasing IL-4 concentration. After virgorous washing, their suppressive capacity on responder $CD4^+CD25^-$ was tested. IL-4 pretreatment did not affect the suppressive capacity of FACS-sorted $CD4^+CD25^{high}$ cells. $1x10^4$ $CD4^+CD25^+$ T_{regs} were added to $5x10^4$ $CD4^+CD25^-$ and $5x10^4$ irradiated PBMCs. The results are representative of three independent experiments.

Effect of IL-4 on T_{regs} in vivo

To verify the striking effect of IL-4 on T_{reg} commitment *in vivo*, we injected IL-4 into normal wild-type B6 mice. We used complexes of recombinant mouse IL-4 (rmIL-4) plus anti-IL-4 monoclonal antibodies (mAb), which have been shown to dramatically increase the potency of the cytokine *in vivo* [313]. Mice injected with rmIL-4 or anti-IL-4 mAb alone did not show signicant changes in the frequency of $CD4^+CD25^+$ cells (Figure 24A-C, E). In contrast the percentage of $CD4^+CD25^+$ T cell dramatically decreased, when the cytokine antibody complexes were injected (Figure 24D, F). Preliminary data confirmed that the decrease of $CD25^+$ T cells reflected a decrease in $Foxp3^+$ T cells, as measured by Foxp3 intracellular staining. This effect was specific for IL-4 and not for the mAb, since another anti-IL-4 mAb also decreased the T_{reg} frequency only in combination with IL-4 (Figure 24 E, F). Upon administration of rmIL-4 plus anti-IL-4 mAb complexes, the total number of $CD4^+CD25^+$ T cell diminished by half (Figure 24G), confirming that the lower percentage was not due to an increase in the $CD4^+CD25^-$ cells, but a real decrease of $CD4^+CD25^+$ cells. In conclusion IL-4 negatively regulates the Treg turnover not only *in vitro* but also *in vivo*.

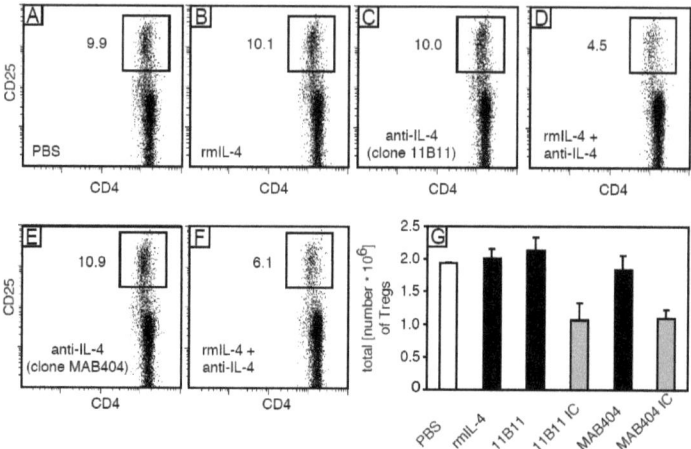

Figure 24: In vivo treatment of mice with IL4 antibody-cytokine complexes.
(A-G) Normal B6 mice were given every other day ip injections of phosphate-buffered saline (PBS), recombinant mouse IL-4 (rmIL-4), anti-IL-4 monoclonal antibody (anti-IL-4 mAb, 11B11 or MAB404), or a mixture of rmIL-4 plus anti-IL-4 mAb (11B11 or MAB404). Mice were analyzed on day 7 by flow cytometry for CD3, CD4 and CD25 expression.
(A-F) Shown is CD25 versus CD4 expression in CD3+ CD4+ spleen cells. Numbers indicate percentages of CD4+ CD25high CD3+ cells.
(G) Total cell counts of CD4+ CD25high cells in spleen from mice in (A-F) are shown as mean ± SD. The data are representative of three independent experiments.

Discussion

The current study reveals that FOXP3-mediated T_{reg} commitment is inhibited by GATA3, which is the key regulator for polarization towards Th2 cells. After differentiation the effector cells become refractory to conversion into a regulatory phenotype and particularly Th2 cells were unable to upregulate FOXP3. In accordance with other studies, we found that CD4$^+$CD25$^-$ were able to upregulate FOXP3 [63,306]. Already committed cells such as memory T cells and Th1 cells showed only moderate and transient FOXP3 induction. Particularily Th2 cells were lacking FOXP3 under all circumstances, which was not caused by endogenous production of IL-4, since IL-4 neutralization did not restore FOXP3 expression. Predominantly naïve T cells could efficiently upregulate FOXP3, suggesting, that FOXP3 plays an important role in the early differentiation process and may act in a similar way as it is known for GATA3 and T-bet, which are essential for commitment of naïve T cells towards Th2 and Th1 respectively. This commitment is characterized by competitive and exclusive expression of these factors [100,314], which we also observed during the differentiation of FOXP3$^+$ iT$_{regs}$, lacking GATA3 or T-bet expression. In this competitive process TGF-β appeared to be mandatory for the induction of FOXP3, possibly by keeping the expression of GATA3 and T-bet low [315,316]. In contrast, differentiating naïve T cells under neutral, "Th0" conditions showed only a transient FOXP3 expression and failed to generate a stable population of FOXP3 expressing cells, but GATA3 and T-bet were upregulated (data not shown). Interestingly, as we and other previously described, FOXP3-inducing factors, such as dexamethasone [262], CTLA-4 [317] and estrogens [263], are also known as inhibitor of GATA3 expression [318-321]. As a result of the differentiation process iT$_{regs}$ evolve as a separate lineage, characterized by its suppressive function and a distinct cytokine and surface receptor profile relative to Th1/Th2 cells.

The successful polarization of naïve T cells towards a Th1 or Th2 immune response is balanced by cytokines. In this context we demonstrated that IL-4 was able to inhibit stable FOXP3 induction mediated by TGF-β and therefore prevented the conversion into the regulatory phenotype. IL-4 has already been shown to negatively regulate the development of naïve T cells into Th1 or the recently-described IL-17 produing T cells (Th17) T cells (Harrington et al., 2005; Pace et al., 2005). Similar effect has been recently described for IL-6, which inhibits combined with TGF-β the generation of iTregs and induced the differentiation into the Th17 cells by an unknown mechanism [323,324]. We hypothesized that the IL-4-dependent mechanism could be mediated by GATA3 and demonstrated that GATA3-inducing IL-4 concentrations were repressing TGF-β-mediated FOXP3 expression, while IL-4 as well as TGF-β signaling were intact. This result suggested a competitive mechanism between GATA3 and FOXP3 transcription factors in determing lineage commitment during the early phase of differentiation. Accordingly we investigated whether GATA3 overexpression affected FOXP3 induction and found that GATA3 overexpressing naïve human T cells were characterized by a reduced capacity to express FOXP3. This inhibitory effect of GATA3 was further confirmed in transgenic mice, expressing GATA3 in T cells (DO11.10:CD2GATA3 transgenic mice). In line with the transient overexpression of GATA3 in human T cells, cells of these mice failed to induce FOXP3 expression upon exposure with antigen in the presence of TGF-β. Strikingly, the DO11.10 CD2GATA3 mice do have peripheral FOXP3$^+$ cells, which however displayed a 10 -25% lower frequency compared to WT DO11.10 mice. Thus GATA3 restrains the development of certain T_{regs} subsets, presumably the inducible, peripheral population and not those of thymic origin. Thymic T cells are undergoing a different maturation process, which may explain the insensitivity of nTreg to GATA3 overexpression [208]. Preliminary results indicate that the repressor of GATA3 (ROG) may play an important role in nTregs, as it has been shown for differentiation into CD8 cells in the thymus [142,325,326]. Interestingly mice lacking GATA3 develop spontaneously into Th1 cells, however it is currently not known, how iT_{regs} develop in these animals (Zhu et al., 2004).

The current study demonstrated that GATA3 represses FOXP3 expression directly by binding to the FOXP3 promoter region. The palindromic GATA-site is located 303 bp upstream of the transcription start site (TSS) in a highly conserved region, which we have previously identified as the FOXP3 promoter [310]. Site-specific mutation of this site increased the activity of promoter constructs, thus revealing the repressive nature of this GATA element. This palindromic GATA element binds GATA3 protein as proven by pull-down experiments. Furthermore, it is shown by chromatin immune precipitation that GATA3 binds this element also in intact cells, indicating that this chromatin region is accessible for GATA3 binding. It is known that GATA3 can induce transcription by chromatin remodelling [110], by directly transactivating promoters [100] or, as shown in the current study, acts as a repressor of gene expression [102,328,329].

The molecular interactions enabling GATA3 to inhibit FOXP3 are not identified yet, but the GATA-binding site is located adjacent to positive, inducing sites, composed of AP-1-NFATc2 sites [310] and GATA3 may compete with the binding of AP-1/NFAT to the promoter (unpublished observations).

The direct inhibition of FOXP3 by this GATA3 dependent mechanism could also affect already existing Tregs, by the same mechanism. However, IL-4 was ineffective to block FOXP3 expression or suppression of already existing Tregs, although IL-4R is expressed and functional on Treg (Pace et al., 2005). This finding underlines our hypothesis that Tregs are commiting as a separate lineage characterized by an imprinted phenotype. Once committed the T cells loose their capacity to convert to another phenotype (Grogan et al., 2001; Szabo et al., 1997; Harrington et al., 2005).

To prove the inhibitory effect of IL-4 on Treg commitment in vivo, we treated mice with IL-4 and anti-IL-4. Only the IL-4/IL-4 mAb complex resulted in a decrease of the amount of Treg (CD25+ and FOXP3$^+$) cells seven days after treatment. This enhancing effect of the antibody on cytokine effect has previously been described for IL-2 and IL-4 [313]. This finding might also explain a study, where anti-IL-4 mAb treatment was interpreted in the sense of IL-4 neutralization instead of IL-4 potentiation [331].

In summary, we demonstrated that GATA3 acts as an inhibitor of FOXP3 expression in early T cell differentiation, by directly binding and repressing the FOXP3 promoter. These data support the idea that Tregs evolve as a separate lineage apart from the Th1 and Th2. These findings will give new perspectives in promoting peripheral tolerance to control autoimmune diseases and allergies on one hand and break tolerance against tumors on the other hand.

Experimental procedures

Mice

Normal C57BL/6 (B6) mice were purchased from the Jackson Laboratories.

Transgenic DO11.10 mice, expressing a T cell receptor for $OVA_{323-339}$ peptide in the context of $H-2^d$, were backcrossed with mice expressing GATA3, driven by the human CD2 locus control region (CD2-GATA3) [308], resulting in DO11.10xCD2-GATA3 mice. Mice used for experiments were backcrossed on a BALB.C background for a minimum of eight generations and used at an age of 8-12 weeks. Mice were housed under specific pathogen-free conditions and all animal studies were performed according to institutional and state guidelines.

Administration of Cytokines and Antibodies In Vivo

Age- and gender-matched normal B6 mice received every other day intraperitoneal (ip) injections of PBS, 1.5 µg rmIL-4, 50 µg anti-IL-4 mAb (11B11 or MAB404), or a mixture of 1.5 µg rmIL-4 plus 50 µg anti-IL-4 mAb (11B11 or MAB404) for 7 days. Thereafter, spleen and LN cells were analyzed by flow cytometry for CD3, CD4 and CD25 expression. The anti-mouse IL-4 mAb MAB404 was obtained from R&D Systems, the second anti-mouse IL-4 mAb 11B11 was purchased from eBioscience.

Isolation of $CD4^+$ T cells

$CD4^+$ T cells were isolated from blood of healthy volunteers using the anti-CD4 magnetic beads (Dynal, Hamburg, Germany) as previously described [269]. The purity of $CD4^+$ T cells was initially tested by FACS and was ≥ 95%.

RNA isolation and cDNA synthesis

RNA was isolated using the RNeasy Mini Kit (Qiagen, Hamburg, Germany) according to the manufacturer's protocol. Reverse transcription of human samples was performed with TaqMan® reverse transcription reagents (Applied Biosystems, Rotkreuz, Switzerland) with random hexamers according to the manufacturer's protocol. For the murine experiments, 100 ng RNA was reverse transcribed using Superscript II (Invitrogen) and random hexmamers (Amersham Biosciences, Roosendaal, The Netherlands) for 50 min. at 42 °C.

Quantitative real-time PCR

The PCR primers and probes detecting human FOXP3 were designed based on the sequences reported in GenBank with the Primer Express software version 1.2 (Applied Biosystems) as follows: EF-1α forward primer and reverse primer as described [332], FOXP3 forward primer 5` GAA ACAG CAC ATT CCC AGA GTT C 3`, FOXP3 reverse primer 5` ATG GCC CAG CGG ATG AG 3`, GATA3 forward primer 5` GCG GGC TCT ATC ACA AAA TGA 3`. The prepared cDNAs were amplified using SYBR®-PCR mastermix (Biorad) according to the recommendations of the manufacturer in an ABI PRISM 7000 Sequence Detection System (Applied Biosystems).

Quantitative PCR of murine samples was performed with Brilliant SYBR Green QPCR master mix (Stratagene, La Jolla, CA, USA) and following primers: Ubiquitin C, 5'- AGGTCAAACAGGAAGACAGACGTA-3' and 5'- TCACACCCAAGAACAAG CACA-3'; Smad-7, 5'-GAAACCGGGGGAACGAAT TAT-3' and 5'- CGCGAGTC TTCTCCTCCCA-3'; TGF-ß$_1$, 5'- TGACGTCACTG GAGTTGTACGG-3' and 5'-GGTTCATGTCATGGATGGTGC-3'. Primer pairs were evaluated for integrity by analysis of the amplification plot, dissociation curves and efficiency of PCR amplification. PCR conditions were 10 min. at 95 °C, followed by 40 cycles of 15 s at 95 °C and 60 °C for 1 min. using an 7300 real-time PCR system (Applied Biosystems). PCR amplification of the housekeeping gene encoding ubiquitin C was performed during each run for each sample to allow normalization between samples. Relative quantification and calculation of the range of confidence was performed using the comparative ΔΔCT method.

Inducible murine Treg culture

Naïve CD4$^+$ T cells (CD4$^+$, CD62L$^+$, CD25$^-$) were isolated from pooled lymph nodes and spleens by FACS-sorting (FACS Aria, BD). 5x10^5 T cells were co-cultured with 2.5x10^4 bone marrow-derived dendritic cells [333] and 0.01 μg/ml
OVA$_{323-339}$ peptide (Ansynth, Roosendaal, The Netherlands) in the presence or absence of 20 ng/μl rhTGF-ß$_1$ (Peprotech, Rocky Hill, NJ, USA) in 48-well plates. After four days, cells were harvested and analyzed for intracellular FOXP3 expression by FACS or gene expression by real-time quantitative RT-PCR.

In vitro T cell differentiation
T cells were stimulated with immobilized plate-bound anti-CD3 (1 µg/ml, Okt3, IgG1) and anti-CD28 (2 µg/ml) in Th1 conditions: 25 ng/ml IL-12, 5 µg/ml anti-IL-4 (R&D systems, Abingdon, UK), in Th2 conditions: 25 ng/ml IL-4, 5 µg/ml anti-IFN-γ, 5 µg/ml anti-IL-12 (R&D systems) or Tregs conditions: 10 ng/ml TGF-β, 5 µg/ml anti-IFN-γ, 5 µg/ml anti-IL-12, 5 µg/ml anti-IL-4. Proliferating cells were expanded in medium containing IL-2 (30 ng/ml).

Western blotting
For FOXP3 analysis on the protein level, 1×10^6 cells $CD4^+CD25^-$ were lysed and loaded next to a protein-mass ladder (Magicmark, Invitrogen) on a NuPAGE 4-12% bis-tris gel (Invitrogen). The proteins were electroblotted onto a PVDF membrane (Amersham Life Science, Dübendorf, Switzerland). Unspecific binding was blocked with BSA and the membranes were subsequently incubated with an 1:200 dilution of goat anti-FOXP3 in blocking buffer (Abcam, Hamburg, Germany) overnight at 4°C. The blots were developed using an anti-goat HRP-labeled mAb (Amersham Biosciences) and visualized with a LAS 1000 camera (Fuji, Urdorf, Switzerland). Membranes were incubated in stripping buffer and re-blocked for 1 h. The membranes were re-probed using anti-GATA3 (HG3-31; Santa Cruz Biotechnology), anti-T-bet (4B10, Santa Cruz Biotechnology, Santa Cruz, CA, USA), anti-GAPDH (6C5, Ambion Ltd, Huntington, United Kingdom), anti-phospho-SMAD2 (138D4), anti-phospho-STAT6 (5A4) and anti-STAT6 (Cell Signaling technology, Allschwil, Switzerland),

Cytokine production asssay
T cells were stimulated with 2×10^{-7} M PMA and 1 µg/ml of ionomycin (Sigma Chemicals, St-Louis, MO, USA) for 4 h. The following mAbs were used: anti-IL-4-PE (8D4-8, BD), anti-IL-13-PE (JES10-5A2, BD Biosciences), anti-IFN-γ-FITC (B27, BD Biosciences). Matched isotype controls were used at the same protein concentration as the respective antibodies. Four-color FACS was performed using an EPICS^{TM} XL-MCL (Beckman Coulter, Nyon, Switzerland) using the software Expo^{TM} 32 version for data acquisition and evaluation.

Flow cytometry

For analysis of FOXP3 expression at the single-cell level, cells were first stained with the monoclonal antibody CD25 (Beckman & Coulter), after fixation and permeabilization, cells were incubated with PE-conjugated monoclonal antibody PCH101 (anti-human FOXP3; eBioscience) based on the manufacturer's recommendations and subjected to FACS (EPICS XL-MCL). For cell surface marker staining, cells were incubated for 20 min at 4 °C in staining buffer with the following antibodies anti–CD152-PE (CTLA-4; BD), or anti–PD-1 (eBiosciences), anti-GITR (R & D Systems, Ltd), anti-CD69 (Beckman & Coulter), anti-CD103 (DakoCytomation, Zug, Switzerland), anti-CD62L (Beckman & Coulter), anti-HLA-DR (Beckman & Coulter). The controls were FITC, PE, or ECD-conjugated mouse IgG1 or rat IgG2a. For staining of mouse cells the following mAbs from BD Biosciences were used following standard techniques as described above: anti-CD3, anti-CD4, anti-CD25. Anti-FcγRII/III antibody (2.4G2, ATCC, Manassas, VA) was included in all stainings to reduce non-specific antibody binding. To isolate naïve murine CD4 T cells from murine DO11.10 or DO11.10xCD2-GATA3 T cells, cells were stained with anti-CD25-FITC, anti-CD62L-PE and anti-CD4-APC prior to sorting. Dead cells were excluded with 4',6-Diamidino-2-phenylindole (DAPI). To analyze murine Foxp3 expression in inducible Treg cultures, cells were stained intracellularly with anti-Foxp3-PE according to manufacturer's instruction, in conjunction with anti-CD4-APC and LIVE/DEAD fixable dead cell stain kit (Invitrogen) to discriminate live cells. All monoclonal antibodies for murine cell stainings were purchased from eBioscience or BD Biosciences.

Cloning of the FOXP3 promoter, construction of mutant constructs

The FOXP3 promoter amplicon was cloned into the pGL3 basic vector (Promega Biotech Inc., Madison, WI, USA) to generate the pGL3 FOXP3 –511/+176. Site-directed mutagenesis in the FOXP3 promoter region, were introduced using the QuickChange kit (Stratagene), according to the manufacturer's instructions. The following primer and its complementary strand were used: GTT TCT CAT GAG CCC TAT TAA GTC ATT CTT ACC TCT CAC CTC TGT GGT GA.

Transfections and reporter gene assays

T cells were rested in serum-free AIM-V medium (Life Technologies, Basel, Switzerland) overnight. An amount of 3.5 µg of the FOXP3 promoter luciferase reporter vector and 0.5 µg phRL-TK was added to 3 $\times 10^6$ CD4$^+$ T cells resuspended in 100 µL of Nucleofector^{TM} solution (Amaxa Biosystems, Cologne, Germany) and electroporated using the U-15 program of the Nucleofector^{TM}. After a 24-hour culture in serum-free conditions and stimuli as indicated in the figures, luciferase activity was measured, by the dual luciferase assay system (Promega Biotech Inc.) according to the manufacturer's instructions. Data were normalized by the activity of renilla luciferase. Jurkat cells were transfected using lipofectamine 2000 (Invitrogen) according to the manufacturer's protocol.

Pull-down Assay

CD4$^+$ T cells were stimulated with PMA and ionomycin for 2 h at 37°C. The cells were pelleted, resuspended in buffer C (20 mM HEPES (pH 7.9), 420 mM NaCl, 1.5 mM MgCl$_2$, 0.2 mM EDTA, 1 mM DTT, protease inhibitors (Sigma, Buchs, Switzerland) and 0.1% NP-40) and lysed on ice for 15 min. Insoluble material was removed by centrifugation. The supernatant was diluted 1:3 with buffer D (as buffer C, but without NaCl). The lysates were incubated with 10 µg of poly(dI-dC; Sigma) and 70 µl of streptavidin-agarose (Amersham Biosciences) carrying biotinylated oligonucleotides, for 3 h at 4 °C. The beads were washed twice with buffer C/D (1:3) and resuspended in DTT-containing loading buffer (NuPAGE; Invitrogen), heated to 70°C for 10 min and the eluants loaded next to a protein-mass ladder (Magicmark, Invitrogen) on a NuPAGE 4-12% bis-tris gel (Invitrogen). The proteins were electroblotted onto a PVDF membrane (Amersham Biosciences) and detected using an anti-GATA3 mAb (Santa Cruz Biotechnology). The blots were developed as described above. Accumulated signals were analyzed using AIDA software (Raytest, Urdorf, Switzerland).

Chromatin Immunoprecipitation

ChIP analysis was performed according to the manufacturer's protocol (Upstate Biotechnology, Inc.) with the following modifications. iTregs and Th2 cells were fixed with 1% formaldehyde for 10 min at room temperature. The chromatin was sheared to 200–1000 bp of length by sonication with five pulses of 10 s at 30% power (Bandelin). The chromatin was pre-cleared for 2 h with normal mouse IgG beads and then incubated with anti-GATA3-agarose beads (HG3-31; Santa Cruz Biotechnology) for 2 h. Washing and elution buffers were used according to the protocol of Upstate Biotechnology. Crosslinks were reversed by incubation at 65 °C for 4 h in the presence of 0.2 M NaCl, and the DNA was purified by phenol/chloroform extraction. The amount of DNA was determined by conventional PCR. The PCR addressed for the FOXP3 promoter region -246 to -511 and was performed using the following primers: 5'-GTG CCC TTT ACG AGT CAT CTG-3' and 5'-GTG CCC TTT ACG AGT CAT CTG-3'. The PCR products were visualized using an ethidium bromide gel.

FACS-sorting of human $CD4^+CD25^+$

PBMC were isolated from buffy coat by density gradient centrifugation over Ficoll/Hypaque gradient. Cells were stained with PE-anti-CD25 and anti-PE magnetic beads (Miltenyi Biotec, Bergisch Gladbach, Germany) and $CD25^+$ cells were enriched using the Midi-MACS system (Miltenyi Biotec). CD25-enriched or -depleted cell populations were stained with FITC-anti-CD4 and sorted into $CD4^+CD25^-$ and $CD4^+CD25^{high}$ on a FACStar Plus (BD Biosciences).

Suppression assay

Samples in triplicate, containing 5×10^4 irradiated PBMCs, 5×10^4 $CD4^+CD25^-$ and 1×10^4 of preactivated or resting $CD4^+CD25^+$ T cells per well were incubated in 96 round-bottom-plates, which were previously coated with 1μg/ml antiCD3 mab or a matched isotype control. Cells were cultured for 4 days, pulsed for the last 10 h with 1 μCi [^{3}H]-thymidine (Hartmann, Braunschweig, Germany) and harvested on glass fiber filters using an automated multisample harvester (LKB, Pharmacia-Wallac, Turku, Finland). Filters were transferred in sample bags with liquid scintillation fluid and analyzed using a β-scintillation counter (Pharmacia-Wallac). Round-bottom 96-well plates were coated with 1 μg/μl anti-CD3 for 1 h at 37 °C and subsequently washed with PBS.

Acknowledgments

This work was supported by the Swiss National Foundation Grants Nr: 31-65436, 3100A0-100164 and 310000-112329, the Ehmann Foundation Chur, the Ernst Goehner Foundation Zug, the Saurer Foundation Zurich and the Swiss Life Zurich.

2.3 Statement of contribution to publications

For the publication titled "Molecular mechanisms underlying FOXP3 induction in human T cells", I have performed all the experiments, except the cell sorting by FACS of the $CD4^+CD25^{high}$ Tregs, which was performed by Beate Rückert (SIAF, Davos, Switzerland).

For the publication titled "GATA3 driven Th2 responses inhibit FOXP3 expression and the formation of regulatory T cells" I have contributed to all figures except of the Figure 19 B and C and Figure 20 A and B, which were performed by Dr. Harmjan Kuipers (Erasmus University, Rotterdam, the Netherlands).
Finally, the experiments for the figure 24 were performed by Dr. Onur Boyman (Scripps Research Institute, La Jolla, USA).

3. Discussion

This thesis revealed mechanisms of FOXP3 gene expression regulation and provides a model of the molecular events underlying Treg cell generation. In a first step the promoter of the human FOXP3 gene was analyzed and led to the identification of NFAT1 and AP1 as inducer of FOXP3 after TCR triggering while GATA3 was discovered to act as a repressor of FOXP3 during differentiation into Th2 cells and in committed Th2 cells.

3.1. Identification and characterization of the human basal FOXP3 promoter

The human promoter was localized by 5'-RACE, at -6221 bp upstream from the translation start site (TSS). The sequence upstream of the UTR shows a high degree of conservation between human, mouse and rat; which indicates that it contains important regulatory elements. The promoter shows cell-specific activity. It was active only in $CD4^+CD25^-$ T cells but not in HELA, HEK or CHO cells.

This region contains a functional TATA box as shown by mutational analysis and EMSA. This TATA box, which is located at –44 bp from TSS, is found in many genes near the transcription start site. The TATA box plays an important role in assembling the transcription machinery at promoters [334]. It is bound by the TATA-binding-protein, which recruits the RNA polymerase II, therefore playing a decisive role in transcription, as it has been shown for other genes such as IL-2 and IL-4 [79,274].

A GC box was identified –141 upstream the transcription start site and was proven to be functional. GC box binding proteins are able to induce basal promoter activity. Sp1 can activate gene expression, whereas Sp3 can repress or activate transcription [277,335-337]. The Sp3-mediated repression has been shown as a result of competition with Sp1 for the binding site [338,339], however Sp3 can also be a strong transcriptional activator. In case of the FOXP3 promoter it appears that Sp1 and Sp3 are not the only proteins to bind to the GC box, since the complex did not fully shift upon treatment with the antibodies against Sp1 and Sp3. Several factors have been identified to bind GC-rich region [273] like Sp4, BTEB1 (basic transcription element binding protein 1) [340], TIEG1 and TIEG2 (TGF-β-inducible early protein genes 1 and 2) [341-343]. Krüppel-like factors do not only bind GC boxes, but also structurally similar CACC-boxes, of which a potential binding site is located near the GC box and may therefore have overlapping binding capacity to Sp1 and Sp3. Sp1 has been shown to regulate ubiquitously expressed housekeeping genes as well as cell type-specific and differentiation-specific gene [344,345]. The promoter region -307/+176 was also weakly active in other cell lines, underlying the importance of other regulatory mechanisms like presence of cell-specific enhancer or silencer. Gene expression might be regulated at the chromatin level as for many genes during differentiation of T cells, such as IL-4 and IFN-γ, in which chromatin opening is an important step in derepressing silenced genes [79,274]. In fact, we found that the chromatin was in a closed and restrictive conformation in non-hematopoietic cells. In $CD4^+CD25^-$ the chromatin was open, but after activation it was demonstrated to become more accessible as measured by histone acetylation. Therefore the human FOXP3 core promoter has a classical structure common to many genes. With a TATA box near the transcription start site at –44 bp and a GC box at –141. These elements are characteristic of housekeeping genes and do not explain the cell-specific expression of FOXP3.

3.2. Cell-specific activity of the FOXP3 promoter

TCR stimulation of human $CD4^+CD25^-$ T cells by an antigen can upregualte FOXP3 expression in previously FOXP3- cells. Interestingly a region of the human FOXP3 promoter seems to be particularly important in maintaining the activity of the promoter and is responsive to TCR stimulation. TCR-induction of the FOXP3 promoter activity was mediated in an AP-1-NFATc2-dependent fashion and was blocked by the addition of the calcineurin inhibitor CsA. The activity of the NFAT proteins is tightly regulated by the calcium/calmodulin-dependent phosphatase calcineurin, which is inhibited by CsA [346]. Calcineurin directly mediates the activation of NFAT molecule by dephosphrylation allowing migration into the nucleus [284]. AP-1 is composed of a heterodimer including Jun and Fos, and is activated through the MAPK kinase pathway [347]. The inhibitors of MAPK kinase pathways only modestly repress the promoter induction as well as mRNA expression in activated $CD4^+CD25^-$ T cells. Interestingly NFAT – AP-1 elements are found in many genes that are inducibly transcribed by TCR-triggering and is part of a common process of T cell activation [284,348-350]. Site-specific mutations deleting these sites were decreasing activity of the promoter. The TCR-mediated induction of FOXP3 promoter activity revealed and confirmed the importance of antigen experience in the thymus and in the periphery for Treg induction.

3.3. Effects of immunosuppressive drugs on FOXP3 expression

The importance of TCR-triggering for generation and activation of Tregs suggests that immunosuppressive drugs might have different effect on the development of tolerance mediated by Tregs. Accordingly CsA inhibited potently the induction of FOXP3 in contrast to rapamycin. CsA is used in the treatment of atopic dermatitis and transplantation and has revolutionized transplant surgery since its introduction in 1983 with a dramatic increase in the survival rate after transplantation [351]. Rapamycin, a macrolide antibiotic produced by Streptomyces hygroscopicus, is a new effective drug used to prevent allograft rejection [352]. However, unlike FK506 and CsA, rapamycin does not inhibit TCR-induced calcineurin activity. Rapamycin exerts its effect by binding to the intracellular immunophilin FK506-binding protein (FKBP12). The rapamycin-FKBP12 complex inhibits the serine/threonine protein kinase called mammalian target of rapamycin (mTOR), whose activation is required for protein synthesis and cell-cycle progression. Therefore, rapamycin blocks signaling in response to cytokines or growth factors, whereas FK506 and CsA exert their inhibitory effects by blocking TCR-induced activation [353]. Consistent with this mechanism of action, it has been shown that rapamycin blocks T-cell-cycle progression from G1 to S phase after activation [353]. Rapamycin promotes TCR-induced anergy even in the presence of costimulation (Figure 25)[354,355].

Interestingly, it has been shown that rapamycin induces proliferation of $CD4^+CD25^+$ T cells, whereas CsA had an inhibitory effect [292]. Therefore it can be concluded that CsA induce tolerance by directly inhibiting activation of effector cells, but at the same time it inhibits the generation of Tregs. The therapeutic usage of CsA may not promote the development of long-term cellular tolerance. Since every T cell has the potential to be activated by TCR, mechanisms that inhibit FOXP3 expression and therefore Treg commitment must exist. TCR triggering is probably only the first step in Treg commitment.

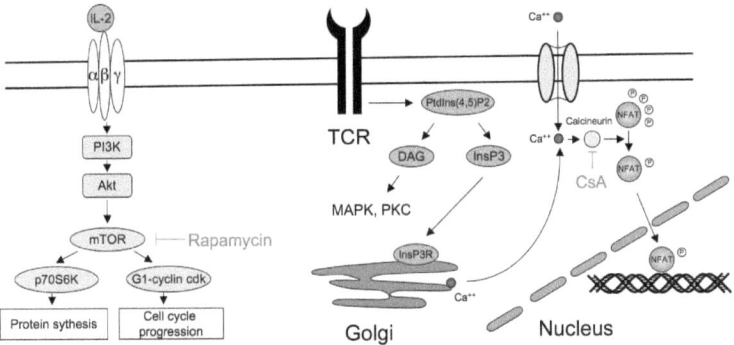

Figure 25: Rapamycin and CsA induce immune suppression by two different pathways. CsA targets calcineurin and NFAT –induction of gene transcription. However, rapamycin inhibits cell cycle progression and protein synthesis. CsA interferes with Treg turnover but not rapamycin.

3.4. Models of Treg response to antigen

Two models, which discriminate between recruitment and generation of Tregs on the site of antigen-triggered immune response, can be proposed to explain the role of FOXP3$^+$ T cells in regulating the immune response. In the first model FOXP3$^+$ T cells are recruited to the site of inflammation and will upon encounter with the antigens regulate the effector antigen-specific T cells and expand to control the intensity of the immune response. In the second model, FOXP3$^-$ cells are recruited and activated by the antigen and give rise to a clonal expansion of a mixture of two populations composed of effectors and FOXP3$^+$ Tregs. In the last model two different options can be hypothetized: already committed effector T cells upon activation can convert to Tregs while keeping their Th1 or Th2 profile. Alternatively, Tregs only differentiate out of naïve T cells as a different lineage (Figure 26).

In order to clarify this point, we analyzed FOXP3 inducibility in different subsets of cells. In accordance with other studies, we found that CD4$^+$CD25$^-$ were able to upregulate FOXP3 [63,306]. Already committed cells such as memory T cells and Th1 cells showed only moderate and transient FOXP3 induction. Interestingly Th2 cells were lacking FOXP3 under all circumstances, which was not caused by endogenous production of IL-4, since IL-4 neutralization did not restore FOXP3 expression.

Predominantly naïve T cells could efficiently upregulate FOXP3, suggesting, that effector Th2 cells lack an intrinsic factor or actively repress FOXP3 expression.

Treg response to the antigen
1. Recruitment to the site of the antigen

2. Generation in contact to antigen

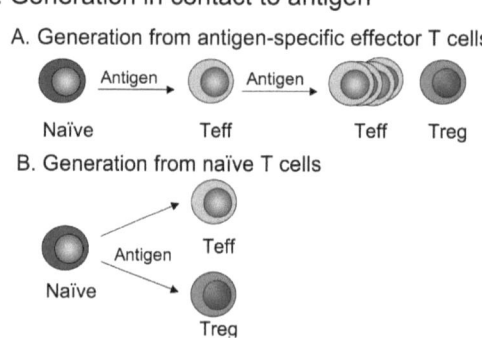

Figure 26: Model of Treg response to the antigen. 1) Differentiated Tregs and effector T cells (Teff) are recruited by chemotaxis to the site of antigen. 2) Generation of Tregs following stimulation by the antigen. A) Upon contact to antigen, antigen-specific T effector cells proliferate and some of them differentiate into Tregs. B) Committed effector T cells lost their capacity to convert to Tregs and Tregs differentiation has to occur out of naïve T cells.

3.5. Differentiating Th2 cells lack FOXP3

The commitment to effector cells is characterized by competitive and exclusive expression of the corresponding transcription factors, GATA3 or T-Bet, while the opposite factor is progressively silenced [100,314]. By using *in vitro* differentiation system, in which naïve T cells are treated with IL-12, IL-4 or TGF-β to become Th1, Th2 or iTreg, respectively, we observed a similar process during the differentiation into FOXP3$^+$ iT$_{regs}$, which were lacking GATA3 or T-bet expression. In this competitive process, TGF-β appeared to be mandatory for the induction of FOXP3, possibly by keeping the expression of GATA3 and T-bet low [315,316,356]. In contrast, differentiating naïve T cells under neutral, "Th0" conditions showed only a transient FOXP3 expression and failed to generate a stable population of FOXP3 expressing cells, but GATA3 and T-bet were upregulated. Interestingly, as we and other previously described, FOXP3-inducing factors, such as dexamethasone [172], CTLA-4 [222] and estrogens [223], are also known as inhibitor of GATA3 expression [224-227]. TGF-β is a molecule with pleiotropic effects on cell proliferation, differentiation, migration, and survival that affect multiple biological processes [357].

TGF-β mediates its biological functions via binding to type I and II transmembrane kinase receptors and phosphorylates intracellular SMAD proteins whose nuclear localization is required for the transcriptional regulation of target genes [358-360]. After association with SMAD4, SMAD2 translocate into the nucleus to bind to SBE (Smad-binding element). SMAD complexes control gene expression by recruiting coactivators that contain histone-acetyl transferase (HAT) activity or histone-deacetylase (HDAC) activity–containing corepressors to activate or repress target genes, respectively [361]. Studies using cells that are deficient in SMAD4 and in the expression of dominant-negative SMADs support the existence of Smad-independent TGF-β signaling pathways [362,363].

Therefore TGF-β might modulate the expression of FOXP3 by different mechanisms. No direct effect of TGF-β was seen on the FOXP3 promoter in transfected CD4+ T cells, indicating that TGF-β might act indirectly by influencing expression of other factors, by binding to other regulatory elements of the FOXP3 genes, or modulate chromatin structure. Interestingly IL-4 was antagonizing TGF-β effect during differentiation of naïve T cells into Tregs. IL-4 has already been shown to negatively regulate the development into Th1 or Th17 cells [65,67].

3.6. GATA3 inhibits FOXP3 expression

GATA3 has been demonstrated to inhibit Th1 commitment during Th2 differentiation. We demonstrated that overexpression of GATA3 inhibited the FOXP3 mRNA expression in naïve CD4 T cells.

Strikingly the DO11.10 CD2GATA3 mice, which overexpress GATA3 under the control of the CD2 Locus Control Region (LCR) have less FOXP3+ cells. In addition, the FOXP3- T cells cannot be induced to express FOXP3. Interestingly, these mice do not develop autoimmunity. This can be explained by homeostatic expansion or other mechanisms are involved in repressing GATA3 activity like ROG [326,364,365] or FOG-1 [366,367] and -2. ROG has been shown to negatively regulate GATA3 activity in CD8 cells. Even when CD8 cells where overexpressing GATA3 they were unable to produce as much IL-4 as CD4 cells do [326]. In addition, ROG has been shown to mediate this effect by sequestering GATA3. In thymocytes GATA3 induces differentiation into CD4. However, when ROG was overexpressed in thymocytes they differentiated into the CD8 lineage [325]. Preliminary data indicate that ROG is expressed by Tregs [142].

Overexpression of GATA3 has already been described to promote airway hyperresponsiveness [302], whereas overexpression of a dominant negative form of this transcription factor [304] or treatment with antisense-induced GATA3 blockade [305] in mice decreased the severity of the disease. In addition, it has been shown in an OVA-tolerance model that co-treatment of mice with cholera toxin, which induces GATA3 expression was able to break tolerance induction [303]. It was thought that GATA3 induces effector functions by stimulating IL-4 and IL-13 secretion. We now describe a new role of GATA3 in promoting the immune response by repressing the generation of FOXP3+ regulatory cells, representing an important checkpoint to avoid cells developping suppressive or anergic phenotype and keep an immune response working until clearance of the pathogen. Under inflammatory conditions the naïve cells "see" the antigen in a pro-inflammatory environment and elimination of antigens is mandatory to avoid damage to the host. Induction of GATA3 might be a mechanism to limit the generation of unwanted Tregs.

GATA3 is probably not the only factor negatively regulating FOXP3 expression, since the GATA3-deficient naïve T cells develop spontaneously into Th1 cells or IFN-γ-secreting cells, under Th2 differentiation conditions [327], rather than becoming Treg. However they only produced a low amount of IFN-γ, which was not comparable with Th1 cells, but this suggests that Treg commitment is not a default pathway. T-bet, the master factor in Th1 cell differentiation might be a candidate, since mutation of a T-bet consensus binding site in the human FOXP3 promoter resulted in an increase of the promoter activity in CD4+ T cells (unpublished data). However, T-bet functions rather as an activator of transcription [314].

3.7. GATA3 represses the FOXP3 promoter activity

We further demonstrated that GATA3 represses FOXP3 expression by direct binding to the promoter. A mutation in the human FOXP3 promoter, disrupting GATA3 binding site was increasing the luciferase activity in transfected CD4+ T cells. The repressor function of GATA3 on FOXP3 promoter activity was further demonstrated by cotransfection of GATA3 with the FOXP3 luciferase construct. GATA3 has been mainly described as a transcription factor, which, induces transcription by chromatin remodelling [110] or directly transactivating promoters [100]. However, its function as a repressor is still unclear, but it has already been described as a repressor of STAT4 [102] and other genes [328,329,368]. The TCR transactivates the FOXP3 promoter in an AP-1-NFATc2-mediated fashion [310] and GATA3 might compete with the binding of NFAT to the promoter at the time of activation (unpublished observations) acting as a repressor by being a weaker inducer than NFATc2-AP-1. In addition, GATA3 might act by recruiting co-repressor complex to the promoter.

3.8. Effect of IL-4 on committed Tregs

Although the IL-4R is expressed and functional on Tregs [322], both induced and naturally occurring Tregs did not convert to effector or FOXP3⁻ T cells upon treatment with IL-4, indicating that after the commitment, the cells have imprinted their phenotype. One explanation might be the expression of an inhibitor of the IL-4 signalling. It has been shown that once committed, the cell looses their capacity to convert easily to another phenotype, although some flexibility exists [109]. Interestingly, we found that the Treg cells were not able to induce GATA3 mRNA as much as the CD25⁻ cells did. Therefore it can be hypothetized that FOXP3 acts as a repressor of GATA3, since loss of FOXP3 expression seems to restore an effector phenotype as shown in the following examples: IPEX patients bearing a mutated FOXP3 and unfunctional FOXP3 have autoreactive FOXP3+ T cells [306], indicating that whitout a functional FOXP3, the Tregs acquired an effector phenotype. In addition, it has been described using cells of Phemphigus Vulgaris (PV) patients, a severe autoimmune bullous skin disorder that Tr1 cells expressing FOXP3 transfected with siRNA against

FOXP3 developed into a Th2 phenotype [369]. Further suggesting that FOXP3 may negatively regulate GATA3 expression.

3.9. Effect of IL-4 on Tregs *in vivo*

In order to analyze the *in vivo* effect of IL-4 on Tregs, we injected a complex of IL-4 and IL-4 antibody to B6 mice every other day during seven days. The number and frequency of $CD4^+CD25^+$ (as well as $CD4^+FOXP3^+$) diminished by half. In these experimental settings, we cannot distinguish whether this drop is caused by a loss of thymic naturally-occuring Tregs or peripherally generated Tregs. According to their turnover in vivo, Tregs are composed of two subsets: one quiescent with long-life span (over 70 days) and the other fraction has a rapid turnover, which is characterized by autoreactive Tregs being continuously activated by tissue self-antigens [370]. Furthermore Tregs have short telomeres indicating that they are highly differentiated cells, which have experienced antigenic stimulation [171] and are long-lived cells. Thus, probably that IL-4 also acts on generation, cells death and conversion of Tregs.

IL-4 and IL-13 have been shown to convert human $CD4^+CD25^-$ naïve T cells into $CD4^+CD25^+$ Tregs *in vitro*. We showed that administration of neutralizing antibodies against IL-4 or IL-13 broke oral tolerance in an antigen-specific manner in a mouse model, with a reduced number of Tregs, suggesting a role for IL-4 and IL-13 in the generation of antigen-specific Tregs during oral tolerance. However in this study FOXP3 was not measured. Since IL-4 has been described as a growth factor for Tregs [322], it should have been distinguished between true generation and proliferation of $CD4^+CD25^-FOXP3^+$ induced by IL-4 [331]. In addition the increase of $CD4^+CD25^+$ in presence of IL-4 was low. As it was described for IL-2 and IL-4 some cytokine neutralizing antibodies *in vitro* have indeed the opposite effect *in vivo*, when injected into mice. The mechanism is unknown but the antibodies might build complexes with the cytokines and thus potentiates its effect [313] making difficult to interpret this result.

The observation that GATA3 represses and avoid activation of the FOXP3 promoter, suggests that at the time of priming the strength of the TCR signaling and the cytokines present in the environment are decisive not only in converting naïve T cells to Th1 or Th2, but also in inducing regulatory T cells. A competition of transcription factors may decide, which direction the immune response will take in order to efficiently fight the antigen. This results from the signals given by the innate immune system and cells, which were previously in contact with the pathogens. Therefore allowing specificity and flexibility to the adaptive system to fight and clear the pathogens.

3.10. Conclusion and outlook

Taken together, we have identified and characterized the human FOXP3 promoter in primary CD4+ T cells. We could explain the molecular mechanisms involved in FOXP3 upregulation following TCR triggering and described that GATA3 acts as an inhibitor of FOXP3 expression that binds directly and represses the FOXP3 promoter (Figure 27). Given the high diversity of T cell subtypes and responses it seems that a hierarchy of mechanisms has evolved during CD4 effector cells development at the transcriptional level to place stringent controls on Th-cell development.

Understanding the molecular mechanisms of Treg induction in the periphery gives new perspectives in generating antigen-specific Treg cells, which can efficiently control autoimmune diseases and allergies. On the other hand inhibiting regulatory T cells [185] in the tumor environment will boost the tumor-immunity. The results of this thesis not only gives new insights into the molecular mechanisms involved into FOXP3 expression and Treg generation, but may also be the starting point for the design of new therapeutical approaches targeting IL-4 in order to modulate Treg development.

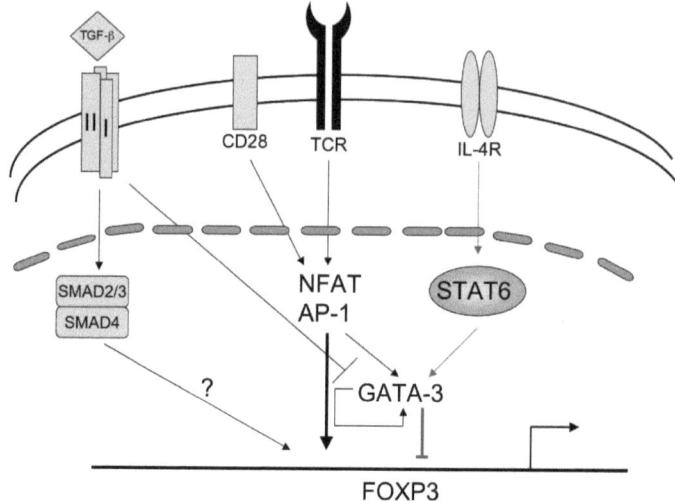

Figure 27: FOXP3 gene regulation in primary T cells. TCR triggering by activating NFAT and AP-1 is a positive regulator of FOXP3 promoter activity, whereas IL-4 by inducing GATA3 expression acts as negative signal.

4. References

1. Diebold, S. S., Kaisho, T., Hemmi, H., Akira, S. & Reis e Sousa, C. Innate antiviral responses by means of TLR7-mediated recognition of single-stranded RNA. *Science* **303**, 1529-1531 (2004).
2. Heil, F. et al. Species-specific recognition of single-stranded RNA via toll-like receptor 7 and 8. *Science* **303**, 1526-1529 (2004).
3. Hornung, V. et al. Replication-dependent potent IFN-alpha induction in human plasmacytoid dendritic cells by a single-stranded RNA virus. *J Immunol* **173**, 5935-5943 (2004).
4. Lund, J., Sato, A., Akira, S., Medzhitov, R. & Iwasaki, A. Toll-like receptor 9-mediated recognition of Herpes simplex virus-2 by plasmacytoid dendritic cells. *J Exp Med* **198**, 513-520 (2003).
5. Hsieh, C. S. et al. Development of TH1 CD4+ T cells through IL-12 produced by Listeria-induced macrophages. *Science* **260**, 547-549 (1993).
6. Gould, I. A., Belok, L. C. & Handwerger, S. Listeria monocytogenes: a rare cause of opportunistic infection in the acquired immunodeficiency syndrome (AIDS) and a new cause of meningitis in AIDS. A case report. *AIDS Res* **2**, 231-234 (1986).
7. Macher, A. M. The pathology of AIDS. *Public Health Rep* **103**, 246-254 (1988).
8. Underhill, D. M. et al. The Toll-like receptor 2 is recruited to macrophage phagosomes and discriminates between pathogens. *Nature* **401**, 811-815 (1999).
9. Poltorak, A. et al. Defective LPS signaling in C3H/HeJ and C57BL/10ScCr mice: mutations in Tlr4 gene. *Science* **282**, 2085-2088 (1998).
10. Platts-Mills, T. A. The role of immunoglobulin E in allergy and asthma. *Am J Respir Crit Care Med* **164**, S1-5 (2001).
11. Tao, X., Constant, S., Jorritsma, P. & Bottomly, K. Strength of TCR signal determines the costimulatory requirements for Th1 and Th2 CD4+ T cell differentiation. *J Immunol* **159**, 5956-5963 (1997).
12. Badou, A. et al. Weak TCR stimulation induces a calcium signal that triggers IL-4 synthesis, stronger TCR stimulation induces MAP kinases that control IFN-gamma production. *Eur J Immunol* **31**, 2487-2496 (2001).
13. Schwartz, R. H. T cell anergy. *Annu Rev Immunol* **21**, 305-334 (2003).

14. Abbas, A. K. The control of T cell activation vs. tolerance. *Autoimmun Rev* **2**, 115-118 (2003).
15. Carreno, B. M. & Collins, M. The B7 family of ligands and its receptors: new pathways for costimulation and inhibition of immune responses. *Annu Rev Immunol* **20**, 29-53 (2002).
16. Chen, L. Co-inhibitory molecules of the B7-CD28 family in the control of T-cell immunity. *Nat Rev Immunol* **4**, 336-347 (2004).
17. Chikuma, S. & Bluestone, J. A. CTLA-4 and tolerance: the biochemical point of view. *Immunol Res* **28**, 241-253 (2003).
18. Coyle, A. J. & Gutierrez-Ramos, J. C. More negative feedback? *Nat Immunol* **4**, 647-648 (2003).
19. Khoury, S. J. & Sayegh, M. H. The roles of the new negative T cell costimulatory pathways in regulating autoimmunity. *Immunity* **20**, 529-538 (2004).
20. Sharpe, A. H. & Freeman, G. J. The B7-CD28 superfamily. *Nat Rev Immunol* **2**, 116-126 (2002).
21. Trinchieri, G. Interleukin-12 and its role in the generation of TH1 cells. *Immunol Today* **14**, 335-338 (1993).
22. Aggarwal, S., Ghilardi, N., Xie, M. H., de Sauvage, F. J. & Gurney, A. L. Interleukin-23 promotes a distinct CD4 T cell activation state characterized by the production of interleukin-17. *J Biol Chem* **278**, 1910-1914 (2003).
23. Pflanz, S. et al. IL-27, a heterodimeric cytokine composed of EBI3 and p28 protein, induces proliferation of naive CD4(+) T cells. *Immunity* **16**, 779-790 (2002).
24. Takeda, A. et al. Cutting edge: role of IL-27/WSX-1 signaling for induction of T-bet through activation of STAT1 during initial Th1 commitment. *J Immunol* **170**, 4886-4890 (2003).
25. Hunter, C. A. New IL-12-family members: IL-23 and IL-27, cytokines with divergent functions. *Nat Rev Immunol* **5**, 521-531 (2005).
26. Gollob, J. A. et al. Altered interleukin-12 responsiveness in Th1 and Th2 cells is associated with the differential activation of STAT5 and STAT1. *Blood* **91**, 1341-1354 (1998).

27. Afkarian, M. et al. T-bet is a STAT1-induced regulator of IL-12R expression in naive CD4+ T cells. *Nat Immunol* **3**, 549-557 (2002).
28. Bacon, C. M. et al. Interleukin 12 induces tyrosine phosphorylation and activation of STAT4 in human lymphocytes. *Proc Natl Acad Sci U S A* **92**, 7307-7311 (1995).
29. Mullen, A. C. et al. Role of T-bet in commitment of TH1 cells before IL-12-dependent selection. *Science* **292**, 1907-1910 (2001).
30. Barbulescu, K. et al. IL-12 and IL-18 differentially regulate the transcriptional activity of the human IFN-gamma promoter in primary CD4+ T lymphocytes. *J Immunol* **160**, 3642-3647 (1998).
31. Amsen, D. et al. Instruction of distinct CD4 T helper cell fates by different notch ligands on antigen-presenting cells. *Cell* **117**, 515-526 (2004).
32. Maekawa, Y. et al. Delta1-Notch3 interactions bias the functional differentiation of activated CD4+ T cells. *Immunity* **19**, 549-559 (2003).
33. Maillard, I., Fang, T. & Pear, W. S. Regulation of lymphoid development, differentiation, and function by the Notch pathway. *Annu Rev Immunol* **23**, 945-974 (2005).
34. Minter, L. M. et al. Inhibitors of gamma-secretase block in vivo and in vitro T helper type 1 polarization by preventing Notch upregulation of Tbx21. *Nat Immunol* **6**, 680-688 (2005).
35. Tanigaki, K. et al. Notch-RBP-J signaling is involved in cell fate determination of marginal zone B cells. *Nat Immunol* **3**, 443-450 (2002).
36. Tanigaki, K. et al. Regulation of alphabeta/gammadelta T cell lineage commitment and peripheral T cell responses by Notch/RBP-J signaling. *Immunity* **20**, 611-622 (2004).
37. Steinbrink, K., Wolfl, M., Jonuleit, H., Knop, J. & Enk, A. H. Induction of tolerance by IL-10-treated dendritic cells. *J Immunol* **159**, 4772-4780 (1997).
38. Kapsenberg, M. L. Dendritic-cell control of pathogen-driven T-cell polarization. *Nat Rev Immunol* **3**, 984-993 (2003).
39. Groux, H., Fournier, N. & Cottrez, F. Role of dendritic cells in the generation of regulatory T cells. *Semin Immunol* **16**, 99-106 (2004).
40. Wakkach, A. et al. Characterization of dendritic cells that induce tolerance and T regulatory 1 cell differentiation in vivo. *Immunity* **18**, 605-617 (2003).

41. Urban, B. C. & Roberts, D. J. Malaria, monocytes, macrophages and myeloid dendritic cells: sticking of infected erythrocytes switches off host cells. *Curr Opin Immunol* **14**, 458-465 (2002).
42. Walther, M. et al. Upregulation of TGF-beta, FOXP3, and CD4+CD25+ regulatory T cells correlates with more rapid parasite growth in human malaria infection. *Immunity* **23**, 287-296 (2005).
43. Tailleux, L. et al. DC-SIGN is the major Mycobacterium tuberculosis receptor on human dendritic cells. *J Exp Med* **197**, 121-127 (2003).
44. Geijtenbeek, T. B. et al. Mycobacteria target DC-SIGN to suppress dendritic cell function. *J Exp Med* **197**, 7-17 (2003).
45. Dolganiuc, A. et al. Hepatitis C virus core and nonstructural protein 3 proteins induce pro- and anti-inflammatory cytokines and inhibit dendritic cell differentiation. *J Immunol* **170**, 5615-5624 (2003).
46. Salio, M., Cella, M., Suter, M. & Lanzavecchia, A. Inhibition of dendritic cell maturation by herpes simplex virus. *Eur J Immunol* **29**, 3245-3253 (1999).
47. Moutaftsi, M., Mehl, A. M., Borysiewicz, L. K. & Tabi, Z. Human cytomegalovirus inhibits maturation and impairs function of monocyte-derived dendritic cells. *Blood* **99**, 2913-2921 (2002).
48. van der Kleij, D. et al. A novel host-parasite lipid cross-talk. Schistosomal lyso-phosphatidylserine activates toll-like receptor 2 and affects immune polarization. *J Biol Chem* **277**, 48122-48129 (2002).
49. McGuirk, P., McCann, C. & Mills, K. H. Pathogen-specific T regulatory 1 cells induced in the respiratory tract by a bacterial molecule that stimulates interleukin 10 production by dendritic cells: a novel strategy for evasion of protective T helper type 1 responses by Bordetella pertussis. *J Exp Med* **195**, 221-231 (2002).
50. Mosmann, T. R., Cherwinski, H., Bond, M. W., Giedlin, M. A. & Coffman, R. L. Two types of murine helper T cell clone. I. Definition according to profiles of lymphokine activities and secreted proteins. *J Immunol* **136**, 2348-2357 (1986).
51. Mosmann, T. R. & Coffman, R. L. TH1 and TH2 cells: different patterns of lymphokine secretion lead to different functional properties. *Annu Rev Immunol* **7**, 145-173 (1989).

52. Coffman, R. L. Origins of the T(H)1-T(H)2 model: a personal perspective. *Nat Immunol* **7**, 539-541 (2006).
53. Heinzel, F. P., Sadick, M. D., Holaday, B. J., Coffman, R. L. & Locksley, R. M. Reciprocal expression of interferon gamma or interleukin 4 during the resolution or progression of murine leishmaniasis. Evidence for expansion of distinct helper T cell subsets. *J Exp Med* **169**, 59-72 (1989).
54. Locksley, R. M. et al. Susceptibility to infectious diseases: Leishmania as a paradigm. *J Infect Dis* **179 Suppl 2**, S305-8 (1999).
55. Sacks, D. & Noben-Trauth, N. The immunology of susceptibility and resistance to Leishmania major in mice. *Nat Rev Immunol* **2**, 845-858 (2002).
56. Riedler, J. et al. Exposure to farming in early life and development of asthma and allergy: a cross-sectional survey. *Lancet* **358**, 1129-1133 (2001).
57. Dahl, M. E., Dabbagh, K., Liggitt, D., Kim, S. & Lewis, D. B. Viral-induced T helper type 1 responses enhance allergic disease by effects on lung dendritic cells. *Nat Immunol* **5**, 337-343 (2004).
58. Weiner, H. L. Induction and mechanism of action of transforming growth factor-beta-secreting Th3 regulatory cells. *Immunol Rev* **182**, 207-214 (2001).
59. Bellinghausen, I., Klostermann, B., Knop, J. & Saloga, J. Human CD4+CD25+ T cells derived from the majority of atopic donors are able to suppress TH1 and TH2 cytokine production. *J Allergy Clin Immunol* **111**, 862-868 (2003).
60. Stassen, M. et al. Differential regulatory capacity of CD25+ T regulatory cells and preactivated CD25+ T regulatory cells on development, functional activation, and proliferation of Th2 cells. *J Immunol* **173**, 267-274 (2004).
61. McKee, A. S. & Pearce, E. J. CD25+CD4+ cells contribute to Th2 polarization during helminth infection by suppressing Th1 response development. *J Immunol* **173**, 1224-1231 (2004).
62. Chinen, J. & Shearer, W. T. Advances in asthma, allergy and immunology series 2004: basic and clinical immunology. *J Allergy Clin Immunol* **114**, 398-405 (2004).
63. Walker, M. R. et al. Induction of FoxP3 and acquisition of T regulatory activity by stimulated human CD4+CD25- T cells. *J Clin Invest* **112**, 1437-1443 (2003).

64. Fantini, M. C. et al. Cutting edge: TGF-beta induces a regulatory phenotype in CD4+CD25- T cells through Foxp3 induction and down-regulation of Smad7. *J Immunol* **172**, 5149-5153 (2004).
65. Park, H. et al. A distinct lineage of CD4 T cells regulates tissue inflammation by producing interleukin 17. *Nat Immunol* **6**, 1133-1141 (2005).
66. Mangan, P. R. et al. Transforming growth factor-beta induces development of the T(H)17 lineage. *Nature* **441**, 231-234 (2006).
67. Harrington, L. E. et al. Interleukin 17-producing CD4+ effector T cells develop via a lineage distinct from the T helper type 1 and 2 lineages. *Nat Immunol* **6**, 1123-1132 (2005).
68. Vanaudenaerde, B. M. et al. The role of interleukin-17 during acute rejection after lung transplantation. *Eur Respir J* **27**, 779-787 (2006).
69. Koenders, M. I. et al. Interleukin-17 Acts Independently of TNF-{alpha} under Arthritic Conditions. *J Immunol* **176**, 6262-6269 (2006).
70. Yssel, H. & Groux, H. Characterization of T cell subpopulations involved in the pathogenesis of asthma and allergic diseases. *Int Arch Allergy Immunol* **121**, 10-18 (2000).
71. Singh, V. K., Mehrotra, S. & Agarwal, S. S. The paradigm of Th1 and Th2 cytokines: its relevance to autoimmunity and allergy. *Immunol Res* **20**, 147-161 (1999).
72. Shinkai, K., Mohrs, M. & Locksley, R. M. Helper T cells regulate type-2 innate immunity in vivo. *Nature* **420**, 825-829 (2002).
73. Voehringer, D., Shinkai, K. & Locksley, R. M. Type 2 immunity reflects orchestrated recruitment of cells committed to IL-4 production. *Immunity* **20**, 267-277 (2004).
74. Yamane, H., Zhu, J. & Paul, W. E. Independent roles for IL-2 and GATA-3 in stimulating naive CD4+ T cells to generate a Th2-inducing cytokine environment. *J Exp Med* **202**, 793-804 (2005).
75. Avni, O. et al. T(H) cell differentiation is accompanied by dynamic changes in histone acetylation of cytokine genes. *Nat Immunol* **3**, 643-651 (2002).
76. van Leeuwen, B. H., Martinson, M. E., Webb, G. C. & Young, I. G. Molecular organization of the cytokine gene cluster, involving the human IL-3, IL-4, IL-5, and GM-CSF genes, on human chromosome 5. *Blood* **73**, 1142-1148 (1989).

77. Mosley, B. et al. The murine interleukin-4 receptor: molecular cloning and characterization of secreted and membrane bound forms. *Cell* **59**, 335-348 (1989).
78. Abdel-Razzak, Z., Garlatti, M., Aggerbeck, M. & Barouki, R. Determination of interleukin-4-responsive region in the human cytochrome P450 2E1 gene promoter. *Biochem Pharmacol* **68**, 1371-1381 (2004).
79. Li-Weber, M. & Krammer, P. H. Regulation of IL4 gene expression by T cells and therapeutic perspectives. *Nat Rev Immunol* **3**, 534-543 (2003).
80. Hou, J. et al. An interleukin-4-induced transcription factor: IL-4 Stat. *Science* **265**, 1701-1706 (1994).
81. Shimoda, K. et al. Lack of IL-4-induced Th2 response and IgE class switching in mice with disrupted Stat6 gene. *Nature* **380**, 630-633 (1996).
82. Kopf, M. et al. Disruption of the murine IL-4 gene blocks Th2 cytokine responses. *Nature* **362**, 245-248 (1993).
83. Kuhn, R., Rajewsky, K. & Muller, W. Generation and analysis of interleukin-4 deficient mice. *Science* **254**, 707-710 (1991).
84. Takeda, K., Kamanaka, M., Tanaka, T., Kishimoto, T. & Akira, S. Impaired IL-13-mediated functions of macrophages in STAT6-deficient mice. *J Immunol* **157**, 3220-3222 (1996).
85. Kurata, H., Lee, H. J., O'Garra, A. & Arai, N. Ectopic expression of activated Stat6 induces the expression of Th2-specific cytokines and transcription factors in developing Th1 cells. *Immunity* **11**, 677-688 (1999).
86. Kuo, C. T. & Leiden, J. M. Transcriptional regulation of T lymphocyte development and function. *Annu Rev Immunol* **17**, 149-187 (1999).
87. Das, J. et al. A critical role for NF-kappa B in GATA3 expression and TH2 differentiation in allergic airway inflammation. *Nat Immunol* **2**, 45-50 (2001).
88. Cannons, J. L. et al. SAP regulates T(H)2 differentiation and PKC-theta-mediated activation of NF-kappaB1. *Immunity* **21**, 693-706 (2004).
89. Shoemaker, J., Saraiva, M. & O'Garra, A. GATA-3 directly remodels the IL-10 locus independently of IL-4 in CD4+ T cells. *J Immunol* **176**, 3470-3479 (2006).
90. Yamashita, M. et al. Essential role of GATA3 for the maintenance of type 2 helper T (Th2) cytokine production and chromatin remodeling at the Th2 cytokine gene loci. *J Biol Chem* **279**, 26983-26990 (2004).

91. Lee, D. U., Agarwal, S. & Rao, A. Th2 lineage commitment and efficient IL-4 production involves extended demethylation of the IL-4 gene. *Immunity* **16**, 649-660 (2002).
92. Lee, G. R., Fields, P. E. & Flavell, R. A. Regulation of IL-4 gene expression by distal regulatory elements and GATA-3 at the chromatin level. *Immunity* **14**, 447-459 (2001).
93. Hutchins, A. S. et al. Gene silencing quantitatively controls the function of a developmental trans-activator. *Mol Cell* **10**, 81-91 (2002).
94. Feng, Q. & Zhang, Y. The MeCP1 complex represses transcription through preferential binding, remodeling, and deacetylating methylated nucleosomes. *Genes Dev* **15**, 827-832 (2001).
95. Makar, K. W. et al. Active recruitment of DNA methyltransferases regulates interleukin 4 in thymocytes and T cells. *Nat Immunol* **4**, 1183-1190 (2003).
96. Siegel, M. D., Zhang, D. H., Ray, P. & Ray, A. Activation of the interleukin-5 promoter by cAMP in murine EL-4 cells requires the GATA-3 and CLE0 elements. *J Biol Chem* **270**, 24548-24555 (1995).
97. Masuda, A., Yoshikai, Y., Kume, H. & Matsuguchi, T. The interaction between GATA proteins and activator protein-1 promotes the transcription of IL-13 in mast cells. *J Immunol* **173**, 5564-5573 (2004).
98. Ranganath, S. et al. GATA-3-dependent enhancer activity in IL-4 gene regulation. *J Immunol* **161**, 3822-3826 (1998).
99. Lee, G. R., Fields, P. E., Griffin, T. J. & Flavell, R. A. Regulation of the Th2 cytokine locus by a locus control region. *Immunity* **19**, 145-153 (2003).
100. Zheng, W. & Flavell, R. A. The transcription factor GATA-3 is necessary and sufficient for Th2 cytokine gene expression in CD4 T cells. *Cell* **89**, 587-596 (1997).
101. Ouyang, W. et al. Inhibition of Th1 development mediated by GATA-3 through an IL-4-independent mechanism. *Immunity* **9**, 745-755 (1998).
102. Usui, T., Nishikomori, R., Kitani, A. & Strober, W. GATA-3 suppresses Th1 development by downregulation of Stat4 and not through effects on IL-12Rbeta2 chain or T-bet. *Immunity* **18**, 415-428 (2003).

103. Ho, I. C., Hodge, M. R., Rooney, J. W. & Glimcher, L. H. The proto-oncogene c-maf is responsible for tissue-specific expression of interleukin-4. *Cell* **85**, 973-983 (1996).
104. Ho, I. C., Lo, D. & Glimcher, L. H. c-maf promotes T helper cell type 2 (Th2) and attenuates Th1 differentiation by both interleukin 4-dependent and -independent mechanisms. *J Exp Med* **188**, 1859-1866 (1998).
105. Li, B., Tournier, C., Davis, R. J. & Flavell, R. A. Regulation of IL-4 expression by the transcription factor JunB during T helper cell differentiation. *EMBO J* **18**, 420-432 (1999).
106. Rincon, M., Derijard, B., Chow, C. W., Davis, R. J. & Flavell, R. A. Reprogramming the signalling requirement for AP-1 (activator protein-1) activation during differentiation of precursor CD4+ T-cells into effector Th1 and Th2 cells. *Genes Funct* **1**, 51-68 (1997).
107. Kim, J. I., Ho, I. C., Grusby, M. J. & Glimcher, L. H. The transcription factor c-Maf controls the production of interleukin-4 but not other Th2 cytokines. *Immunity* **10**, 745-751 (1999).
108. Ouyang, W. et al. Stat6-independent GATA-3 autoactivation directs IL-4-independent Th2 development and commitment. *Immunity* **12**, 27-37 (2000).
109. Messi, M. et al. Memory and flexibility of cytokine gene expression as separable properties of human T(H)1 and T(H)2 lymphocytes. *Nat Immunol* **4**, 78-86 (2003).
110. Lee, H. J. et al. GATA-3 induces T helper cell type 2 (Th2) cytokine expression and chromatin remodeling in committed Th1 cells. *J Exp Med* **192**, 105-115 (2000).
111. Hogquist, K. A., Baldwin, T. A. & Jameson, S. C. Central tolerance: learning self-control in the thymus. *Nat Rev Immunol* **5**, 772-782 (2005).
112. Bosselut, R. CD4/CD8-lineage differentiation in the thymus: from nuclear effectors to membrane signals. *Nat Rev Immunol* **4**, 529-540 (2004).
113. Strominger, J. L. Developmental biology of T cell receptors. *Science* **244**, 943-950 (1989).
114. Jameson, S. C., Hogquist, K. A. & Bevan, M. J. Positive selection of thymocytes. *Annu Rev Immunol* **13**, 93-126 (1995).

115. Gallegos, A. M. & Bevan, M. J. Central tolerance to tissue-specific antigens mediated by direct and indirect antigen presentation. *J Exp Med* **200**, 1039-1049 (2004).
116. Mathis, D. & Benoist, C. Back to central tolerance. *Immunity* **20**, 509-516 (2004).
117. Zehn, D. & Bevan, M. J. T Cells with Low Avidity for a Tissue-Restricted Antigen Routinely Evade Central and Peripheral Tolerance and Cause Autoimmunity. *Immunity* (2006).
118. Liu, G. Y. et al. Low avidity recognition of self-antigen by T cells permits escape from central tolerance. *Immunity* **3**, 407-415 (1995).
119. Perez, V. L. et al. Induction of peripheral T cell tolerance in vivo requires CTLA-4 engagement. *Immunity* **6**, 411-417 (1997).
120. Schwartz, R. H. A cell culture model for T lymphocyte clonal anergy. *Science* **248**, 1349-1356 (1990).
121. Russell, J. H. Activation-induced death of mature T cells in the regulation of immune responses. *Curr Opin Immunol* **7**, 382-388 (1995).
122. Nossal, G. J. Tolerance and ways to break it. *Ann N Y Acad Sci* **690**, 34-41 (1993).
123. O'neill, E. J. et al. Natural and induced regulatory T cells. *Ann N Y Acad Sci* **1029**, 180-192 (2004).
124. Shevach, E. M. Regulatory T cells in autoimmmunity*. *Annu Rev Immunol* **18**, 423-449 (2000).
125. Gershon, R. K. & Kondo, K. Cell interactions in the induction of tolerance: the role of thymic lymphocytes. *Immunology* **18**, 723-737 (1970).
126. Gershon, R. K. & Kondo, K. Infectious immunological tolerance. *Immunology* **21**, 903-914 (1971).
127. Roncarolo, M. G., Bacchetta, R., Bordignon, C., Narula, S. & Levings, M. K. Type 1 T regulatory cells. *Immunol Rev* **182**, 68-79 (2001).
128. Inobe, J. et al. IL-4 is a differentiation factor for transforming growth factor-beta secreting Th3 cells and oral administration of IL-4 enhances oral tolerance in experimental allergic encephalomyelitis. *Eur J Immunol* **28**, 2780-2790 (1998).
129. Mayer, L. & Shao, L. Therapeutic potential of oral tolerance. *Nat Rev Immunol* **4**, 407-419 (2004).

130. Wardrop, R. M. r. & Whitacre, C. C. Oral tolerance in the treatment of inflammatory autoimmune diseases. *Inflamm Res* **48**, 106-119 (1999).
131. Thompson, H. S. & Staines, N. A. Gastric administration of type II collagen delays the onset and severity of collagen-induced arthritis in rats. *Clin Exp Immunol* **64**, 581-586 (1986).
132. Zhang, Z. J., Davidson, L., Eisenbarth, G. & Weiner, H. L. Suppression of diabetes in nonobese diabetic mice by oral administration of porcine insulin. *Proc Natl Acad Sci U S A* **88**, 10252-10256 (1991).
133. Bergerot, I., Fabien, N., Maguer, V. & Thivolet, C. Oral administration of human insulin to NOD mice generates CD4+ T cells that suppress adoptive transfer of diabetes. *J Autoimmun* **7**, 655-663 (1994).
134. Sayegh, M. H. et al. Down-regulation of the immune response to histocompatibility antigens and prevention of sensitization by skin allografts by orally administered alloantigen. *Transplantation* **53**, 163-166 (1992).
135. Hancock, W. W., Sayegh, M. H., Kwok, C. A., Weiner, H. L. & Carpenter, C. B. Oral, but not intravenous, alloantigen prevents accelerated allograft rejection by selective intragraft Th2 cell activation. *Transplantation* **55**, 1112-1118 (1993).
136. Eigenmann, P. A. Future therapeutic options in food allergy. *Allergy* **58**, 1217-1223 (2003).
137. Bitar, D. M. & Whitacre, C. C. Suppression of experimental autoimmune encephalomyelitis by the oral administration of myelin basic protein. *Cell Immunol* **112**, 364-370 (1988).
138. Higgins, P. J. & Weiner, H. L. Suppression of experimental autoimmune encephalomyelitis by oral administration of myelin basic protein and its fragments. *J Immunol* **140**, 440-445 (1988).
139. Chen, Y., Kuchroo, V. K., Inobe, J., Hafler, D. A. & Weiner, H. L. Regulatory T cell clones induced by oral tolerance: suppression of autoimmune encephalomyelitis. *Science* **265**, 1237-1240 (1994).
140. Fukaura, H. et al. Induction of circulating myelin basic protein and proteolipid protein-specific transforming growth factor-beta1-secreting Th3 T cells by oral administration of myelin in multiple sclerosis patients. *J Clin Invest* **98**, 70-77 (1996).

141. Groux, H. et al. A CD4+ T-cell subset inhibits antigen-specific T-cell responses and prevents colitis. *Nature* **389**, 737-742 (1997).
142. Cobbold, S. P. et al. Regulatory T cells and dendritic cells in transplantation tolerance: molecular markers and mechanisms. *Immunol Rev* **196**, 109-124 (2003).
143. Waldmann, H. et al. Regulatory T cells and organ transplantation. *Semin Immunol* **16**, 119-126 (2004).
144. Akdis, M. et al. Immune responses in healthy and allergic individuals are characterized by a fine balance between allergen-specific T regulatory 1 and T helper 2 cells. *J Exp Med* **199**, 1567-1575 (2004).
145. Shevach, E. M. CD4+ CD25+ suppressor T cells: more questions than answers. *Nat Rev Immunol* **2**, 389-400 (2002).
146. Takahashi, T. et al. Immunologic self-tolerance maintained by CD25+CD4+ naturally anergic and suppressive T cells: induction of autoimmune disease by breaking their anergic/suppressive state. *Int Immunol* **10**, 1969-1980 (1998).
147. Read, S., Malmstrom, V. & Powrie, F. Cytotoxic T lymphocyte-associated antigen 4 plays an essential role in the function of CD25(+)CD4(+) regulatory cells that control intestinal inflammation. *J Exp Med* **192**, 295-302 (2000).
148. Shimizu, J., Yamazaki, S., Takahashi, T., Ishida, Y. & Sakaguchi, S. Stimulation of CD25(+)CD4(+) regulatory T cells through GITR breaks immunological self-tolerance. *Nat Immunol* **3**, 135-142 (2002).
149. McHugh, R. S. et al. CD4(+)CD25(+) immunoregulatory T cells: gene expression analysis reveals a functional role for the glucocorticoid-induced TNF receptor. *Immunity* **16**, 311-323 (2002).
150. Sandner, S. E. et al. Role of the programmed death-1 pathway in regulation of alloimmune responses in vivo. *J Immunol* **174**, 3408-3415 (2005).
151. Annacker, O. et al. Essential role for CD103 in the T cell-mediated regulation of experimental colitis. *J Exp Med* **202**, 1051-1061 (2005).
152. Huehn, J. et al. Developmental stage, phenotype, and migration distinguish naive- and effector/memory-like CD4+ regulatory T cells. *J Exp Med* **199**, 303-313 (2004).

153. Baecher-Allan, C., Brown, J. A., Freeman, G. J. & Hafler, D. A. CD4+CD25high regulatory cells in human peripheral blood. *J Immunol* **167**, 1245-1253 (2001).

154. Cao, D. et al. Isolation and functional characterization of regulatory CD25brightCD4+ T cells from the target organ of patients with rheumatoid arthritis. *Eur J Immunol* **33**, 215-223 (2003).

155. Sakaguchi, S., Sakaguchi, N., Asano, M., Itoh, M. & Toda, M. Immunologic self-tolerance maintained by activated T cells expressing IL-2 receptor alpha-chains (CD25). Breakdown of a single mechanism of self-tolerance causes various autoimmune diseases. *J Immunol* **155**, 1151-1164 (1995).

156. Suri-Payer, E., Amar, A. Z., Thornton, A. M. & Shevach, E. M. CD4+CD25+ T cells inhibit both the induction and effector function of autoreactive T cells and represent a unique lineage of immunoregulatory cells. *J Immunol* **160**, 1212-1218 (1998).

157. Itoh, M. et al. Thymus and autoimmunity: production of CD25+CD4+ naturally anergic and suppressive T cells as a key function of the thymus in maintaining immunologic self-tolerance. *J Immunol* **162**, 5317-5326 (1999).

158. Salomon, B. et al. B7/CD28 costimulation is essential for the homeostasis of the CD4+CD25+ immunoregulatory T cells that control autoimmune diabetes. *Immunity* **12**, 431-440 (2000).

159. Takahashi, T. et al. Immunologic self-tolerance maintained by CD25(+)CD4(+) regulatory T cells constitutively expressing cytotoxic T lymphocyte-associated antigen 4. *J Exp Med* **192**, 303-310 (2000).

160. Tung, K. S., Garza, K. M., Lou, Y. & Bagavant, H. Autoimmune ovarian disease: mechanism of induction and prevention. *J Soc Gynecol Investig* **8**, S49-51 (2001).

161. Strobel, P., Preisshofen, T., Helmreich, M., Muller-Hermelink, H. K. & Marx, A. Pathomechanisms of paraneoplastic myasthenia gravis. *Clin Dev Immunol* **10**, 7-12 (2003).

162. Kriegel, M. A. et al. Defective suppressor function of human CD4+ CD25+ regulatory T cells in autoimmune polyglandular syndrome type II. *J Exp Med* **199**, 1285-1291 (2004).

163. Ehrenstein, M. R. et al. Compromised function of regulatory T cells in rheumatoid arthritis and reversal by anti-TNFalpha therapy. *J Exp Med* **200**, 277-285 (2004).

164. Kipnis, J., Avidan, H., Caspi, R. R. & Schwartz, M. Dual effect of CD4+CD25+ regulatory T cells in neurodegeneration: a dialogue with microglia. *Proc Natl Acad Sci U S A* **101 Suppl 2**, 14663-14669 (2004).

165. Haas, J. et al. Reduced suppressive effect of CD4+CD25high regulatory T cells on the T cell immune response against myelin oligodendrocyte glycoprotein in patients with multiple sclerosis. *Eur J Immunol* **35**, 3343-3352 (2005).

166. Beyer, M. et al. Reduced frequencies and suppressive function of CD4+CD25hi regulatory T cells in patients with chronic lymphocytic leukemia after therapy with fludarabine. *Blood* **106**, 2018-2025 (2005).

167. Liu, R. et al. Cooperation of invariant NKT cells and CD4+CD25+ T regulatory cells in the prevention of autoimmune myasthenia. *J Immunol* **175**, 7898-7904 (2005).

168. Longhi, M. S. et al. Functional study of CD4+CD25+ regulatory T cells in health and autoimmune hepatitis. *J Immunol* **176**, 4484-4491 (2006).

169. Pop, S. M., Wong, C. P., Culton, D. A., Clarke, S. H. & Tisch, R. Single cell analysis shows decreasing FoxP3 and TGFbeta1 coexpressing CD4+CD25+ regulatory T cells during autoimmune diabetes. *J Exp Med* **201**, 1333-1346 (2005).

170. Aruna, B. V., Sela, M. & Mozes, E. Suppression of myasthenogenic responses of a T cell line by a dual altered peptide ligand by induction of CD4+CD25+ regulatory cells. *Proc Natl Acad Sci U S A* **102**, 10285-10290 (2005).

171. Taams, L. S. et al. Antigen-specific T cell suppression by human CD4+CD25+ regulatory T cells. *Eur J Immunol* **32**, 1621-1630 (2002).

172. Jarvinen, L. Z., Blazar, B. R., Adeyi, O. A., Strom, T. B. & Noelle, R. J. CD154 on the surface of CD4+CD25+ regulatory T cells contributes to skin transplant tolerance. *Transplantation* **76**, 1375-1379 (2003).

173. Taylor, P. A., Lees, C. J. & Blazar, B. R. The infusion of ex vivo activated and expanded CD4(+)CD25(+) immune regulatory cells inhibits graft-versus-host disease lethality. *Blood* **99**, 3493-3499 (2002).

174. O'Garra, A. & Vieira, P. Regulatory T cells and mechanisms of immune system control. *Nat Med* **10**, 801-805 (2004).
175. Hoffmann, P. & Edinger, M. CD4+CD25+ regulatory T cells and graft-versus-host disease. *Semin Hematol* **43**, 62-69 (2006).
176. Hoffmann, P., Eder, R., Kunz-Schughart, L. A., Andreesen, R. & Edinger, M. Large-scale in vitro expansion of polyclonal human CD4(+)CD25high regulatory T cells. *Blood* **104**, 895-903 (2004).
177. Liyanage, U. K. et al. Prevalence of regulatory T cells is increased in peripheral blood and tumor microenvironment of patients with pancreas or breast adenocarcinoma. *J Immunol* **169**, 2756-2761 (2002).
178. Somasundaram, R. et al. Inhibition of cytolytic T lymphocyte proliferation by autologous CD4+/CD25+ regulatory T cells in a colorectal carcinoma patient is mediated by transforming growth factor-beta. *Cancer Res* **62**, 5267-5272 (2002).
179. Wolf, A. M. et al. Increase of regulatory T cells in the peripheral blood of cancer patients. *Clin Cancer Res* **9**, 606-612 (2003).
180. Ormandy, L. A. et al. Increased populations of regulatory T cells in peripheral blood of patients with hepatocellular carcinoma. *Cancer Res* **65**, 2457-2464 (2005).
181. Karube, K. et al. Expression of FoxP3, a key molecule in CD4CD25 regulatory T cells, in adult T-cell leukaemia/lymphoma cells. *Br J Haematol* **126**, 81-84 (2004).
182. Marshall, N. A. et al. Immunosuppressive regulatory T cells are abundant in the reactive lymphocytes of Hodgkin lymphoma. *Blood* **103**, 1755-1762 (2004).
183. Viguier, M. et al. Foxp3 expressing CD4+CD25(high) regulatory T cells are overrepresented in human metastatic melanoma lymph nodes and inhibit the function of infiltrating T cells. *J Immunol* **173**, 1444-1453 (2004).
184. Gray, C. P., Arosio, P. & Hersey, P. Association of increased levels of heavy-chain ferritin with increased CD4+ CD25+ regulatory T-cell levels in patients with melanoma. *Clin Cancer Res* **9**, 2551-2559 (2003).
185. Curiel, T. J. et al. Specific recruitment of regulatory T cells in ovarian carcinoma fosters immune privilege and predicts reduced survival. *Nat Med* **10**, 942-949 (2004).

186. Curiel, T. J. et al. Specific recruitment of regulatory T cells in ovarian carcinoma fosters immune privilege and predicts reduced survival. *Nat Med* (2004).
187. Onizuka, S. et al. Tumor rejection by in vivo administration of anti-CD25 (interleukin-2 receptor alpha) monoclonal antibody. *Cancer Res* **59**, 3128-3133 (1999).
188. Shimizu, J., Yamazaki, S. & Sakaguchi, S. Induction of tumor immunity by removing CD25+CD4+ T cells: a common basis between tumor immunity and autoimmunity. *J Immunol* **163**, 5211-5218 (1999).
189. Nakamura, K., Kitani, A. & Strober, W. Cell contact-dependent immunosuppression by CD4(+)CD25(+) regulatory T cells is mediated by cell surface-bound transforming growth factor beta. *J Exp Med* **194**, 629-644 (2001).
190. Ostroukhova, M. et al. Treg-mediated immunosuppression involves activation of the Notch-HES1 axis by membrane-bound TGF-beta. *J Clin Invest* **116**, 996-1004 (2006).
191. Ostroukhova, M. et al. Tolerance induced by inhaled antigen involves CD4(+) T cells expressing membrane-bound TGF-beta and FOXP3. *J Clin Invest* **114**, 28-38 (2004).
192. Thornton, A. M. & Shevach, E. M. CD4+CD25+ immunoregulatory T cells suppress polyclonal T cell activation in vitro by inhibiting interleukin 2 production. *J Exp Med* **188**, 287-296 (1998).
193. Lucas, P. J., Kim, S. J., Melby, S. J. & Gress, R. E. Disruption of T cell homeostasis in mice expressing a T cell-specific dominant negative transforming growth factor beta II receptor. *J Exp Med* **191**, 1187-1196 (2000).
194. Kulkarni, A. B. et al. Transforming growth factor beta 1 null mutation in mice causes excessive inflammatory response and early death. *Proc Natl Acad Sci U S A* **90**, 770-774 (1993).
195. Grossman, W. J. et al. Human T regulatory cells can use the perforin pathway to cause autologous target cell death. *Immunity* **21**, 589-601 (2004).
196. Hsieh, C. S., Zheng, Y., Liang, Y., Fontenot, J. D. & Rudensky, A. Y. An intersection between the self-reactive regulatory and nonregulatory T cell receptor repertoires. *Nat Immunol* **7**, 401-410 (2006).

197. Stephens, L. A. & Mason, D. CD25 is a marker for CD4+ thymocytes that prevent autoimmune diabetes in rats, but peripheral T cells with this function are found in both CD25+ and CD25- subpopulations. *J Immunol* **165**, 3105-3110 (2000).
198. Papiernik, M., de Moraes, M. L., Pontoux, C., Vasseur, F. & Penit, C. Regulatory CD4 T cells: expression of IL-2R alpha chain, resistance to clonal deletion and IL-2 dependency. *Int Immunol* **10**, 371-378 (1998).
199. Bensinger, S. J., Bandeira, A., Jordan, M. S., Caton, A. J. & Laufer, T. M. Major histocompatibility complex class II-positive cortical epithelium mediates the selection of CD4(+)25(+) immunoregulatory T cells. *J Exp Med* **194**, 427-438 (2001).
200. Walker, L. S., Chodos, A., Eggena, M., Dooms, H. & Abbas, A. K. Antigen-dependent proliferation of CD4+ CD25+ regulatory T cells in vivo. *J Exp Med* **198**, 249-258 (2003).
201. Jordan, M. S. et al. Thymic selection of CD4+CD25+ regulatory T cells induced by an agonist self-peptide. *Nat Immunol* **2**, 301-306 (2001).
202. Apostolou, I., Sarukhan, A., Klein, L. & von Boehmer, H. Origin of regulatory T cells with known specificity for antigen. *Nat Immunol* **3**, 756-763 (2002).
203. Tai, X., Cowan, M., Feigenbaum, L. & Singer, A. CD28 costimulation of developing thymocytes induces Foxp3 expression and regulatory T cell differentiation independently of interleukin 2. *Nat Immunol* **6**, 152-162 (2005).
204. Taguchi, O. et al. Tissue-specific suppressor T cells involved in self-tolerance are activated extrathymically by self-antigens. *Immunology* **82**, 365-369 (1994).
205. Hsieh, C. S. et al. Recognition of the Peripheral Self by Naturally Arising CD25(+) CD4(+) T Cell Receptors. *Immunity* **21**, 267-277 (2004).
206. Fontenot, J. D. & Rudensky, A. Y. A well adapted regulatory contrivance: regulatory T cell development and the forkhead family transcription factor Foxp3. *Nat Immunol* **6**, 331-337 (2005).
207. Seddon, B. & Mason, D. Peripheral autoantigen induces regulatory T cells that prevent autoimmunity. *J Exp Med* **189**, 877-882 (1999).
208. Watanabe, N. et al. Hassall's corpuscles instruct dendritic cells to induce CD4+CD25+ regulatory T cells in human thymus. *Nature* **436**, 1181-1185 (2005).

209. Liang, S. et al. Conversion of CD4+ CD25- cells into CD4+ CD25+ regulatory T cells in vivo requires B7 costimulation, but not the thymus. *J Exp Med* **201**, 127-137 (2005).
210. Chen, W. et al. Conversion of peripheral CD4+CD25- naive T cells to CD4+CD25+ regulatory T cells by TGF-beta induction of transcription factor Foxp3. *J Exp Med* **198**, 1875-1886 (2003).
211. Jiang, Q. et al. Cell biology of IL-7, a key lymphotrophin. *Cytokine Growth Factor Rev* **16**, 513-533 (2005).
212. von Freeden-Jeffry, U. et al. Lymphopenia in interleukin (IL)-7 gene-deleted mice identifies IL-7 as a nonredundant cytokine. *J Exp Med* **181**, 1519-1526 (1995).
213. Peschon, J. J. et al. Early lymphocyte expansion is severely impaired in interleukin 7 receptor-deficient mice. *J Exp Med* **180**, 1955-1960 (1994).
214. Khaled, A. R. & Durum, S. K. Lymphocide: cytokines and the control of lymphoid homeostasis. *Nat Rev Immunol* **2**, 817-830 (2002).
215. Peffault de Latour, R. et al. Ontogeny, function and peripheral homeostasis of regulatory T cells in the absence of Interleukin-7. *Blood* (2006).
216. Rosenberg, S. A. et al. IL-7 administration to humans leads to expansion of CD8+ and CD4+ cells but a relative decrease of CD4+ T-regulatory cells. *J Immunother* **29**, 313-319 (2006).
217. Liu, W. et al. CD127 expression inversely correlates with FoxP3 and suppressive function of human CD4(+) T reg cells. *J Exp Med* **203**, 1701-1711 (2006).
218. Seddiki, N. et al. Expression of interleukin (IL)-2 and IL-7 receptors discriminates between human regulatory and activated T cells. *J Exp Med* **203**, 1693-1700 (2006).
219. Zhang, H. et al. Lymphopenia and interleukin-2 therapy alter homeostasis of CD4+CD25+ regulatory T cells. *Nat Med* **11**, 1238-1243 (2005).
220. Fontenot, J. D., Gavin, M. A. & Rudensky, A. Y. Foxp3 programs the development and function of CD4+CD25+ regulatory T cells. *Nat Immunol* **4**, 330-336 (2003).

221. D'Cruz, L. M. & Klein, L. Development and function of agonist-induced CD25+Foxp3+ regulatory T cells in the absence of interleukin 2 signaling. *Nat Immunol* **6**, 1152-1159 (2005).
222. Thornton, A. M., Donovan, E. E., Piccirillo, C. A. & Shevach, E. M. Cutting edge: IL-2 is critically required for the in vitro activation of CD4+CD25+ T cell suppressor function. *J Immunol* **172**, 6519-6523 (2004).
223. de la Rosa, M., Rutz, S., Dorninger, H. & Scheffold, A. Interleukin-2 is essential for CD4+CD25+ regulatory T cell function. *Eur J Immunol* **34**, 2480-2488 (2004).
224. Setoguchi, R., Hori, S., Takahashi, T. & Sakaguchi, S. Homeostatic maintenance of natural Foxp3(+) CD25(+) CD4(+) regulatory T cells by interleukin (IL)-2 and induction of autoimmune disease by IL-2 neutralization. *J Exp Med* **201**, 723-735 (2005).
225. Frank, D. A., Robertson, M. J., Bonni, A., Ritz, J. & Greenberg, M. E. Interleukin 2 signaling involves the phosphorylation of Stat proteins. *Proc Natl Acad Sci U S A* **92**, 7779-7783 (1995).
226. Zorn, E. et al. IL-2 regulates FOXP3 expression in human CD4+CD25+ regulatory T cells through a STAT dependent mechanism and induces the expansion of these cells in vivo. *Blood* (2006).
227. Antov, A., Yang, L., Vig, M., Baltimore, D. & Van Parijs, L. Essential role for STAT5 signaling in CD25+CD4+ regulatory T cell homeostasis and the maintenance of self-tolerance. *J Immunol* **171**, 3435-3441 (2003).
228. Snow, J. W. et al. Loss of tolerance and autoimmunity affecting multiple organs in STAT5A/5B-deficient mice. *J Immunol* **171**, 5042-5050 (2003).
229. Fantini, M. C. et al. Transforming growth factor beta induced FoxP3+ regulatory T cells suppress Th1 mediated experimental colitis. *Gut* **55**, 671-680 (2006).
230. Rao, P. E., Petrone, A. L. & Ponath, P. D. Differentiation and expansion of T cells with regulatory function from human peripheral lymphocytes by stimulation in the presence of TGF-{beta}. *J Immunol* **174**, 1446-1455 (2005).
231. Peng, Y., Laouar, Y., Li, M. O., Green, E. A. & Flavell, R. A. TGF-beta regulates in vivo expansion of Foxp3-expressing CD4+CD25+ regulatory T

cells responsible for protection against diabetes. *Proc Natl Acad Sci U S A* **101**, 4572-4577 (2004).

232. Marie, J. C., Letterio, J. J., Gavin, M. & Rudensky, A. Y. TGF-beta1 maintains suppressor function and Foxp3 expression in CD4+CD25+ regulatory T cells. *J Exp Med* **201**, 1061-1067 (2005).

233. Kretschmer, K. et al. Inducing and expanding regulatory T cell populations by foreign antigen. *Nat Immunol* **6**, 1219-1227 (2005).

234. Brunkow, M. E. et al. Disruption of a new forkhead/winged-helix protein, scurfin, results in the fatal lymphoproliferative disorder of the scurfy mouse. *Nat Genet* **27**, 68-73 (2001).

235. Godfrey, V. L., Rouse, B. T. & Wilkinson, J. E. Transplantation of T cell-mediated, lymphoreticular disease from the scurfy (sf) mouse. *Am J Pathol* **145**, 281-286 (1994).

236. Waterhouse, P. et al. Lymphoproliferative disorders with early lethality in mice deficient in Ctla-4. *Science* **270**, 985-988 (1995).

237. Tivol, E. A. et al. Loss of CTLA-4 leads to massive lymphoproliferation and fatal multiorgan tissue destruction, revealing a critical negative regulatory role of CTLA-4. *Immunity* **3**, 541-547 (1995).

238. Dieckmann, D., Plottner, H., Berchtold, S., Berger, T. & Schuler, G. Ex vivo isolation and characterization of CD4(+)CD25(+) T cells with regulatory properties from human blood. *J Exp Med* **193**, 1303-1310 (2001).

239. Chatenoud, L., Salomon, B. & Bluestone, J. A. Suppressor T cells--they're back and critical for regulation of autoimmunity!. *Immunol Rev* **182**, 149-163 (2001).

240. Hori, S., Nomura, T. & Sakaguchi, S. Control of regulatory T cell development by the transcription factor Foxp3. *Science* **299**, 1057-1061 (2003).

241. Yagi, H. et al. Crucial role of FOXP3 in the development and function of human CD25+CD4+ regulatory T cells. *Int Immunol* **16**, 1643-1656 (2004).

242. Allan, S. E. et al. The role of 2 FOXP3 isoforms in the generation of human CD4+ Tregs. *J Clin Invest* **115**, 3276-3284 (2005).

243. Feuerer, M., Benoist, C. & Mathis, D. Green T(R) cells. *Immunity* **22**, 271-272 (2005).

244. Fontenot, J. D. et al. Regulatory T cell lineage specification by the forkhead transcription factor foxp3. *Immunity* **22**, 329-341 (2005).

245. Wan, Y. Y. & Flavell, R. A. Identifying Foxp3-expressing suppressor T cells with a bicistronic reporter. *Proc Natl Acad Sci U S A* **102**, 5126-5131 (2005).
246. Kaestner, K. H., Knochel, W. & Martinez, D. E. Unified nomenclature for the winged helix/forkhead transcription factors. *Genes Dev* **14**, 142-146 (2000).
247. Wu, Y. et al. FOXP3 Controls Regulatory T Cell Function through Cooperation with NFAT. *Cell* **126**, 375-387 (2006).
248. Khattri, R., Cox, T., Yasayko, S. A. & Ramsdell, F. An essential role for Scurfin in CD4+CD25+ T regulatory cells. *Nat Immunol* **4**, 337-342 (2003).
249. Kasprowicz, D. J. et al. Dynamic regulation of FoxP3 expression controls the balance between CD4(+) T cell activation and cell death. *Eur J Immunol* **35**, 3424-3432 (2005).
250. Khattri, R. et al. The amount of scurfin protein determines peripheral T cell number and responsiveness. *J Immunol* **167**, 6312-6320 (2001).
251. Miller, J. F. & Heath, W. R. Self-ignorance in the peripheral T-cell pool. *Immunol Rev* **133**, 131-150 (1993).
252. Carballido, J. M., Carballido-Perrig, N., Oberli-Schrammli, A., Heusser, C. H. & Blaser, K. Regulation of IgE and IgG4 responses by allergen specific T-cell clones to bee venom phospholipase A2 in vitro. *J Allergy Clin Immunol* **93**, 758-767 (1994).
253. Sakaguchi, S., Fukuma, K., Kuribayashi, K. & Masuda, T. Organ-specific autoimmune diseases induced in mice by elimination of T cell subset. I. Evidence for the active participation of T cells in natural self-tolerance; deficit of a T cell subset as a possible cause of autoimmune disease. *J Exp Med* **161**, 72-87 (1985).
254. Shevach, E. M. Certified professionals: CD4(+)CD25(+) suppressor T cells. *J Exp Med* **193**, F41-6 (2001).
255. Thornton, A. M. & Shevach, E. M. Suppressor effector function of CD4+CD25+ immunoregulatory T cells is antigen nonspecific. *J Immunol* **164**, 183-190 (2000).
256. Dittmer, U. et al. Functional impairment of CD8(+) T cells by regulatory T cells during persistent retroviral infection. *Immunity* **20**, 293-303 (2004).

257. Bennett, C. L. et al. A rare polyadenylation signal mutation of the FOXP3 gene (AAUAAA-->AAUGAA) leads to the IPEX syndrome. *Immunogenetics* **53**, 435-439 (2001).
258. Schubert, L. A., Jeffery, E., Zhang, Y., Ramsdell, F. & Ziegler, S. F. Scurfin (FOXP3) acts as a repressor of transcription and regulates T cell activation. *J Biol Chem* **276**, 37672-37679 (2001).
259. Bettelli, E., Dastrange, M. & Oukka, M. Foxp3 interacts with nuclear factor of activated T cells and NF-kappa B to repress cytokine gene expression and effector functions of T helper cells. *Proc Natl Acad Sci U S A* **102**, 5138-5143 (2005).
260. Levy-Lahad, E. & Wildin, R. S. Neonatal diabetes mellitus, enteropathy, thrombocytopenia, and endocrinopathy: Further evidence for an X-linked lethal syndrome. *J Pediatr* **138**, 577-580 (2001).
261. Godfrey, V. L., Wilkinson, J. E. & Russell, L. B. X-linked lymphoreticular disease in the scurfy (sf) mutant mouse. *Am J Pathol* **138**, 1379-1387 (1991).
262. Karagiannidis, C. et al. Glucocorticoids upregulate FOXP3 expression and regulatory T cells in asthma. *J Allergy Clin Immunol* **114**, 1425-1433 (2004).
263. Polanczyk, M. J. et al. Cutting edge: estrogen drives expansion of the CD4+CD25+ regulatory T cell compartment. *J Immunol* **173**, 2227-2230 (2004).
264. Luo, X. et al. Systemic transforming growth factor-beta1 gene therapy induces Foxp3+ regulatory cells, restores self-tolerance, and facilitates regeneration of beta cell function in overtly diabetic nonobese diabetic mice. *Transplantation* **79**, 1091-1096 (2005).
265. Fu, S. et al. TGF-beta induces Foxp3 + T-regulatory cells from CD4 + CD25 - precursors. *Am J Transplant* **4**, 1614-1627 (2004).
266. Mayor, C. et al. VISTA : visualizing global DNA sequence alignments of arbitrary length. *Bioinformatics* **16**, 1046-1047 (2000).
267. Wohlfahrt, J. G. et al. Ephrin-A1 suppresses Th2 cell activation and provides a regulatory link to lung epithelial cells. *J Immunol* **172**, 843-850 (2004).
268. Hamalainen, H. K. et al. Identification and validation of endogenous reference genes for expression profiling of T helper cell differentiation by quantitative real-time RT-PCR. *Anal Biochem* **299**, 63-70 (2001).

269. Kunzmann, S. et al. SARA and Hgs attenuate susceptibility to TGF-beta1-mediated T cell suppression. *Faseb J* **17**, 194-202 (2003).

270. Schmidt-Weber, C. B., Rao, A. & Lichtman, A. H. Integration of TCR and IL-4 signals through STAT6 and the regulation of IL-4 gene expression. *Mol Immunol* **37**, 767-774 (2000).

271. Hagen, G., Dennig, J., Preiss, A., Beato, M. & Suske, G. Functional analyses of the transcription factor Sp4 reveal properties distinct from Sp1 and Sp3. *J Biol Chem* **270**, 24989-24994 (1995).

272. Nan, X. et al. Transcriptional repression by the methyl-CpG-binding protein MeCP2 involves a histone deacetylase complex. *Nature* **393**, 386-389 (1998).

273. Suske, G. The Sp-family of transcription factors. *Gene* **238**, 291-300 (1999).

274. Lin, J. X. & Leonard, W. J. The immediate-early gene product Egr-1 regulates the human interleukin-2 receptor beta-chain promoter through noncanonical Egr and Sp1 binding sites. *Mol Cell Biol* **17**, 3714-3722 (1997).

275. Agarwal, S. & Rao, A. Modulation of chromatin structure regulates cytokine gene expression during T cell differentiation. *Immunity* **9**, 765-775 (1998).

276. Ramsdell, F. Foxp3 and natural regulatory T cells: key to a cell lineage? *Immunity* **19**, 165-168 (2003).

277. Majello, B., De Luca, P. & Lania, L. Sp3 is a bifunctional transcription regulator with modular independent activation and repression domains. *J Biol Chem* **272**, 4021-4026 (1997).

278. Chen, T. C., Cobbold, S. P., Fairchild, P. J. & Waldmann, H. Generation of anergic and regulatory T cells following prolonged exposure to a harmless antigen. *J Immunol* **172**, 5900-5907 (2004).

279. Polanczyk, M. J., Hopke, C., Huan, J., Vandenbark, A. A. & Offner, H. Enhanced FoxP3 expression and Treg cell function in pregnant and estrogen-treated mice. *J Neuroimmunol* **170**, 85-92 (2005).

280. Polanczyk, M. J., Hopke, C., Vandenbark, A. A. & Offner, H. Estrogen-mediated immunomodulation involves reduced activation of effector T cells, potentiation of treg cells, and enhanced expression of the PD-1 costimulatory pathway. *J Neurosci Res* (2006).

281. Bluestone, J. A. New perspectives of CD28-B7-mediated T cell costimulation. *Immunity* **2**, 555-559 (1995).

282. Abbas, A. K., Murphy, K. M. & Sher, A. Functional diversity of helper T lymphocytes. *Nature* **383**, 787-793 (1996).

283. O'Garra, A. Cytokines induce the development of functionally heterogeneous T helper cell subsets. *Immunity* **8**, 275-283 (1998).

284. Jain, J. et al. The T-cell transcription factor NFATp is a substrate for calcineurin and interacts with Fos and Jun. *Nature* **365**, 352-355 (1993).

285. Jain, J., McCaffrey, P. G., Valge-Archer, V. E. & Rao, A. Nuclear factor of activated T cells contains Fos and Jun. *Nature* **356**, 801-804 (1992).

286. Jain, J., Miner, Z. & Rao, A. Analysis of the preexisting and nuclear forms of nuclear factor of activated T cells. *J Immunol* **151**, 837-848 (1993).

287. Rooney, J. W., Hoey, T. & Glimcher, L. H. Coordinate and cooperative roles for NF-AT and AP-1 in the regulation of the murine IL-4 gene. *Immunity* **2**, 473-483 (1995).

288. Dolganov, G. et al. Coexpression of the interleukin-13 and interleukin-4 genes correlates with their physical linkage in the cytokine gene cluster on human chromosome 5q23-31. *Blood* **87**, 3316-3326 (1996).

289. Perkins, D. et al. Regulation of CTLA-4 expression during T cell activation. *J Immunol* **156**, 4154-4159 (1996).

290. Dunn, C. J., Wagstaff, A. J., Perry, C. M., Plosker, G. L. & Goa, K. L. Cyclosporin: an updated review of the pharmacokinetic properties, clinical efficacy and tolerability of a microemulsion-based formulation (neoral)1 in organ transplantation. *Drugs* **61**, 1957-2016 (2001).

291. Baan, C. C. et al. Differential effect of calcineurin inhibitors, anti-CD25 antibodies and rapamycin on the induction of FOXP3 in human T cells. *Transplantation* **80**, 110-117 (2005).

292. Battaglia, M., Stabilini, A. & Roncarolo, M. G. Rapamycin selectively expands CD4+CD25+FoxP3+ regulatory T cells. *Blood* **105**, 4743-4748 (2005).

293. Asai, K. et al. T cell hyporesponsiveness induced by oral administration of ovalbumin is associated with impaired NFAT nuclear translocation and p27kip1 degradation. *J Immunol* **169**, 4723-4731 (2002).

294. Heissmeyer, V. et al. Calcineurin imposes T cell unresponsiveness through targeted proteolysis of signaling proteins. *Nat Immunol* **5**, 255-265 (2004).

295. Gavin, M. A., Clarke, S. R., Negrou, E., Gallegos, A. & Rudensky, A. Homeostasis and anergy of CD4(+)CD25(+) suppressor T cells in vivo. *Nat Immunol* **3**, 33-41 (2002).
296. Romagnani, S. The Th1/Th2 paradigm. *Immunol Today* **18**, 263-266 (1997).
297. Szabo, S. J., Dighe, A. S., Gubler, U. & Murphy, K. M. Regulation of the interleukin (IL)-12R beta 2 subunit expression in developing T helper 1 (Th1) and Th2 cells. *J Exp Med* **185**, 817-824 (1997).
298. Rogge, L. et al. Selective expression of an interleukin-12 receptor component by human T helper 1 cells. *Journal of Experimental Medicine* **185**, 825-831 (1997).
299. Zhu, J., Yamane, H., Cote-Sierra, J., Guo, L. & Paul, W. E. GATA-3 promotes Th2 responses through three different mechanisms: induction of Th2 cytokine production, selective growth of Th2 cells and inhibition of Th1 cell-specific factors. *Cell Res* **16**, 3-10 (2006).
300. Usui, T. et al. T-bet regulates Th1 responses through essential effects on GATA-3 function rather than on IFNG gene acetylation and transcription. *J Exp Med* **203**, 755-766 (2006).
301. Hwang, E. S., Szabo, S. J., Schwartzberg, P. L. & Glimcher, L. H. T helper cell fate specified by kinase-mediated interaction of T-bet with GATA-3. *Science* **307**, 430-433 (2005).
302. Yamashita, N. et al. Involvement of GATA-3-dependent Th2 lymphocyte activation in airway hyperresponsiveness. *Am J Physiol Lung Cell Mol Physiol* (2006).
303. Oriss, T. B. et al. Dynamics of dendritic cell phenotype and interactions with CD4+ T cells in airway inflammation and tolerance. *J Immunol* **174**, 854-863 (2005).
304. Zhang, D. H. et al. Inhibition of allergic inflammation in a murine model of asthma by expression of a dominant-negative mutant of GATA-3. *Immunity* **11**, 473-482 (1999).
305. Finotto, S. et al. Treatment of allergic airway inflammation and hyperresponsiveness by antisense-induced local blockade of GATA-3 expression. *J Exp Med* **193**, 1247-1260 (2001).

306. Gavin, M. A. et al. Single-cell analysis of normal and FOXP3-mutant human T cells: FOXP3 expression without regulatory T cell development. *Proc Natl Acad Sci U S A* **103**, 6659-6664 (2006).
307. Ostroukhova, M. & Ray, A. CD25+ T cells and regulation of allergen-induced responses. *Curr Allergy Asthma Rep* **5**, 35-41 (2005).
308. Nawijn, M. C. et al. Enforced expression of GATA-3 in transgenic mice inhibits Th1 differentiation and induces the formation of a T1/ST2-expressing Th2-committed T cell compartment in vivo. *J Immunol* **167**, 724-732 (2001).
309. Nakao, A. et al. Identification of Smad7, a TGFbeta-inducible antagonist of TGF-beta signalling. *Nature* **389**, 631-635 (1997).
310. Mantel, P. Y. et al. Molecular mechanisms underlying FOXP3 induction in human T cells. *J Immunol* **176**, 3593-3602 (2006).
311. Lee, H. J. et al. Signals and nuclear factors that regulate the expression of interleukin-4 and interleukin-5 genes in helper T cells. *J Allergy Clin Immunol* **94**, 594-604 (1994).
312. Klein, S. C. et al. An improved, sensitive, non-radioactive in situ hybridization method for the detection of cytokine mRNAs. *APMIS* **103**, 345-353 (1995).
313. Boyman, O., Kovar, M., Rubinstein, M. P., Surh, C. D. & Sprent, J. Selective stimulation of T cell subsets with antibody-cytokine immune complexes. *Science* **311**, 1924-1927 (2006).
314. Szabo, S. J. et al. A novel transcription factor, T-bet, directs Th1 lineage commitment. *Cell* **100**, 655-669 (2000).
315. Gorelik, L., Fields, P. E. & Flavell, R. A. Cutting edge: TGF-beta inhibits Th type 2 development through inhibition of GATA-3 expression. *J Immunol* **165**, 4773-4777 (2000).
316. Heath, V. L., Murphy, E. E., Crain, C., Tomlinson, M. G. & O'Garra, A. TGF-beta1 down-regulates Th2 development and results in decreased IL-4-induced STAT6 activation and GATA-3 expression. *Eur J Immunol* **30**, 2639-2649 (2000).
317. Zheng, S. G. et al. TGF-beta requires CTLA-4 early after T cell activation to induce FoxP3 and generate adaptive CD4+CD25+ regulatory cells. *J Immunol* **176**, 3321-3329 (2006).

318. Bour-Jordan, H. et al. CTLA-4 regulates the requirement for cytokine-induced signals in T(H)2 lineage commitment. *Nat Immunol* **4**, 182-188 (2003).

319. Lambert, K. C. et al. Estrogen receptor alpha (ERalpha) deficiency in macrophages results in increased stimulation of CD4+ T cells while 17beta-estradiol acts through ERalpha to increase IL-4 and GATA-3 expression in CD4+ T cells independent of antigen presentation. *J Immunol* **175**, 5716-5723 (2005).

320. Nasta, F., Ubaldi, V., Pace, L., Doria, G. & Pioli, C. Cytotoxic T-lymphocyte antigen-4 inhibits GATA-3 but not T-bet mRNA expression during T helper cell differentiation. *Immunology* **117**, 358-367 (2006).

321. Quan, A., McCall, M. N. & Sewell, W. A. Dexamethasone inhibits the binding of nuclear factors to the IL-5 promoter in human CD4 T cells. *J Allergy Clin Immunol* **108**, 340-348 (2001).

322. Pace, L., Pioli, C. & Doria, G. IL-4 modulation of CD4+CD25+ T regulatory cell-mediated suppression. *J Immunol* **174**, 7645-7653 (2005).

323. Veldhoen, M., Hocking, R. J., Atkins, C. J., Locksley, R. M. & Stockinger, B. TGFbeta in the context of an inflammatory cytokine milieu supports de novo differentiation of IL-17-producing T cells. *Immunity* **24**, 179-189 (2006).

324. Bettelli, E. et al. Reciprocal developmental pathways for the generation of pathogenic effector T(H)17 and regulatory T cells. *Nature* (2006).

325. Hernandez-Hoyos, G., Anderson, M. K., Wang, C., Rothenberg, E. V. & Alberola-Ila, J. GATA-3 expression is controlled by TCR signals and regulates CD4/CD8 differentiation. *Immunity* **19**, 83-94 (2003).

326. Miaw, S. C., Choi, A., Yu, E., Kishikawa, H. & Ho, I. C. ROG, repressor of GATA, regulates the expression of cytokine genes. *Immunity* **12**, 323-333 (2000).

327. Zhu, J. et al. Conditional deletion of Gata3 shows its essential function in T(H)1-T(H)2 responses. *Nat Immunol* **5**, 1157-1165 (2004).

328. Sadat, M. A. et al. GATA-3 represses gp91phox gene expression in eosinophil-committed HL-60-C15 cells. *FEBS Lett* **436**, 390-394 (1998).

329. Schwenger, G. T. et al. GATA-3 has dual regulatory functions in human interleukin-5 transcription. *J Biol Chem* **276**, 48502-48509 (2001).

330. Grogan, J. L. et al. Early transcription and silencing of cytokine genes underlie polarization of T helper cell subsets. *Immunity* **14**, 205-215 (2001).
331. Skapenko, A., Kalden, J. R., Lipsky, P. E. & Schulze-Koops, H. The IL-4 receptor alpha-chain-binding cytokines, IL-4 and IL-13, induce forkhead box P3-expressing CD25+CD4+ regulatory T cells from CD25-CD4+ precursors. *J Immunol* **175**, 6107-6116 (2005).
332. Karagiannidis, C. et al. Activin A is an acute allergen-responsive cytokine and provides a link to TGF-beta-mediated airway remodeling in asthma. *J Allergy Clin Immunol* **117**, 111-118 (2006).
333. Lutz, M. B. et al. An advanced culture method for generating large quantities of highly pure dendritic cells from mouse bone marrow. *J Immunol Methods* **223**, 77-92 (1999).
334. Benoist, C. & Chambon, P. In vivo sequence requirements of the SV40 early promotor region. *Nature* **290**, 304-310 (1981).
335. Dennig, J., Beato, M. & Suske, G. An inhibitor domain in Sp3 regulates its glutamine-rich activation domains. *Embo J* **15**, 5659-5667 (1996).
336. Sowa, Y. et al. Sp3, but not Sp1, mediates the transcriptional activation of the p21/WAF1/Cip1 gene promoter by histone deacetylase inhibitor. *Cancer Res* **59**, 4266-4270 (1999).
337. Gartel, A. L. et al. Myc represses the p21(WAF1/CIP1) promoter and interacts with Sp1/Sp3. *Proc Natl Acad Sci U S A* **98**, 4510-4515 (2001).
338. Birnbaum, M. J. et al. Sp1 trans-activation of cell cycle regulated promoters is selectively repressed by Sp3. *Biochemistry* **34**, 16503-16508 (1995).
339. Yu, B., Datta, P. K. & Bagchi, S. Stability of the Sp3-DNA complex is promoter-specific: Sp3 efficiently competes with Sp1 for binding to promoters containing multiple Sp-sites. *Nucleic Acids Res* **31**, 5368-5376 (2003).
340. Imataka, H. et al. Two regulatory proteins that bind to the basic transcription element (BTE), a GC box sequence in the promoter region of the rat P-4501A1 gene. *Embo J* **11**, 3663-3671 (1992).
341. Subramaniam, M. et al. Identification of a novel TGF-beta-regulated gene encoding a putative zinc finger protein in human osteoblasts. *Nucleic Acids Res* **23**, 4907-4912 (1995).

342. Cook, T., Gebelein, B., Mesa, K., Mladek, A. & Urrutia, R. Molecular cloning and characterization of TIEG2 reveals a new subfamily of transforming growth factor-beta-inducible Sp1-like zinc finger-encoding genes involved in the regulation of cell growth. *J Biol Chem* **273**, 25929-25936 (1998).

343. Fautsch, M. P. et al. TGFbeta-inducible early gene (TIEG) also codes for early growth response alpha (EGRalpha): evidence of multiple transcripts from alternate promoters. *Genomics* **51**, 408-416 (1998).

344. Chen, H. M., Pahl, H. L., Scheibe, R. J., Zhang, D. E. & Tenen, D. G. The Sp1 transcription factor binds the CD11b promoter specifically in myeloid cells in vivo and is essential for myeloid-specific promoter activity. *J Biol Chem* **268**, 8230-8239 (1993).

345. Zhang, D. E. et al. Sp1 is a critical factor for the monocytic specific expression of human CD14. *J Biol Chem* **269**, 11425-11434 (1994).

346. Schreiber, S. L. & Crabtree, G. R. The mechanism of action of cyclosporin A and FK506. *Immunol Today* **13**, 136-142 (1992).

347. Eferl, R. & Wagner, E. F. AP-1: a double-edged sword in tumorigenesis. *Nat Rev Cancer* **3**, 859-868 (2003).

348. McCaffrey, P. G. et al. Isolation of the cyclosporin-sensitive T cell transcription factor NFATp. *Science* **262**, 750-754 (1993).

349. Northrop, J. P. et al. NF-AT components define a family of transcription factors targeted in T-cell activation. *Nature* **369**, 497-502 (1994).

350. Tokumitsu, H., Masuda, E. S., Tsuboi, A., Arai, K. & Arai, N. Purification of the 120 kDa component of the human nuclear factor of activated T cells (NF-AT): reconstitution of binding activity to the cis-acting element of the GM-CSF and IL-2 promoter with AP-1. *Biochem Biophys Res Commun* **196**, 737-744 (1993).

351. Sigal, N. H. & Dumont, F. J. Cyclosporin A, FK-506, and rapamycin: pharmacologic probes of lymphocyte signal transduction. *Annu Rev Immunol* **10**, 519-560 (1992).

352. Kahan, B. D. & Camardo, J. S. Rapamycin: clinical results and future opportunities. *Transplantation* **72**, 1181-1193 (2001).

353. Abraham, R. T. & Wiederrecht, G. J. Immunopharmacology of rapamycin. *Annu Rev Immunol* **14**, 483-510 (1996).

354. Wiederrecht, G. J. et al. Mechanism of action of rapamycin: new insights into the regulation of G1-phase progression in eukaryotic cells. *Prog Cell Cycle Res* **1**, 53-71 (1995).
355. Powell, J. D., Lerner, C. G. & Schwartz, R. H. Inhibition of cell cycle progression by rapamycin induces T cell clonal anergy even in the presence of costimulation. *J Immunol* **162**, 2775-2784 (1999).
356. Bright, J. J. & Sriram, S. TGF-beta inhibits IL-12-induced activation of Jak-STAT pathway in T lymphocytes. *J Immunol* **161**, 1772-1777 (1998).
357. Blobe, G. C., Schiemann, W. P. & Lodish, H. F. Role of transforming growth factor beta in human disease. *N Engl J Med* **342**, 1350-1358 (2000).
358. Massague, J. TGF-beta signal transduction. *Annu Rev Biochem* **67**, 753-791 (1998).
359. Derynck, R. & Zhang, Y. E. Smad-dependent and Smad-independent pathways in TGF-beta family signalling. *Nature* **425**, 577-584 (2003).
360. Shi, Y. & Massague, J. Mechanisms of TGF-beta signaling from cell membrane to the nucleus. *Cell* **113**, 685-700 (2003).
361. Massague, J. & Wotton, D. Transcriptional control by the TGF-beta/Smad signaling system. *EMBO J* **19**, 1745-1754 (2000).
362. Engel, M. E., McDonnell, M. A., Law, B. K. & Moses, H. L. Interdependent SMAD and JNK signaling in transforming growth factor-beta-mediated transcription. *J Biol Chem* **274**, 37413-37420 (1999).
363. Yu, L., Hebert, M. C. & Zhang, Y. E. TGF-beta receptor-activated p38 MAP kinase mediates Smad-independent TGF-beta responses. *EMBO J* **21**, 3749-3759 (2002).
364. Mariani, L., Lohning, M., Radbruch, A. & Hofer, T. Transcriptional control networks of cell differentiation: insights from helper T lymphocytes. *Prog Biophys Mol Biol* **86**, 45-76 (2004).
365. Kang, B. Y., Miaw, S. C. & Ho, I. C. ROG negatively regulates T-cell activation but is dispensable for Th-cell differentiation. *Mol Cell Biol* **25**, 554-562 (2005).
366. Zhou, M. et al. Friend of GATA-1 represses GATA-3-dependent activity in CD4+ T cells. *J Exp Med* **194**, 1461-1471 (2001).

367. Kurata, H. et al. Friend of GATA is expressed in naive Th cells and functions as a repressor of GATA-3-mediated Th2 cell development. *J Immunol* **168**, 4538-4545 (2002).
368. Schwenger, G. T., Mordvinov, V. A. & Sanderson, C. J. Transcription factor GATA-3 is involved in repression of promoter activity of the human interleukin-4 gene. *Biochemistry (Mosc)* **70**, 1065-1069 (2005).
369. Veldman, C. et al. Inhibition of the transcription factor Foxp3 converts desmoglein 3-specific type 1 regulatory T cells into Th2-like cells. *J Immunol* **176**, 3215-3222 (2006).
370. Fisson, S. et al. Continuous activation of autoreactive CD4+ CD25+ regulatory T cells in the steady state. *J Exp Med* **198**, 737-746 (2003).

Pour mes Parents

ACKNOWLEDGEMENTS

I express my special thanks to Dr. Carsten B. Schmidt-Weber, who supervised my research at the Swiss Institute of Allergy and Asthma Research (SIAF).

I would like to thank Prof. Dr. Kurt Blaser, director of the Swiss Institute of Allergy and Asthma Research (SIAF), for the opportunity to write a PhD thesis at the SIAF.

I express my gratitude to Prof. Dr. Roland Wenger for accepting the responsibility for my PhD thesis and for the critical review of this thesis.

I thank all the present and former members of the Asthma group at the SIAF, especially Steffen Kunzmann, Christian Karagiannidis, Nadia Ouaked, Kerstin Siegmund, Claudio Bassin and Beate Rückert for the practical help during work and the good times, both within and outside the lab.

My appreciation goes out to all other co-workers of the Swiss Institute of Allergy and Asthma Research (SIAF), who supported me in theoretical and practical aspects of the thesis and provided a good working atmosphere.

I thank my parents for encouraging me and supporting me mentally and materially.

Die VDM Verlagsservicegesellschaft sucht für wissenschaftliche Verlage abgeschlossene und herausragende

Dissertationen, Habilitationen, Diplomarbeiten, Master Theses, Magisterarbeiten usw.

für die kostenlose Publikation als Fachbuch.

Sie verfügen über eine Arbeit, die hohen inhaltlichen und formalen Ansprüchen genügt, und haben Interesse an einer honorarvergüteten Publikation?

Dann senden Sie bitte erste Informationen über sich und Ihre Arbeit per Email an *info@vdm-vsg.de*.

Sie erhalten kurzfristig unser Feedback!

VDM Verlagsservicegesellschaft mbH
Dudweiler Landstr. 99   Telefon +49 681 3720 174
D - 66123 Saarbrücken   Fax   +49 681 3720 1749
www.vdm-vsg.de

Die VDM Verlagsservicegesellschaft mbH vertritt

Printed by Books on Demand GmbH, Norderstedt / Germany